# A Checklist of Indian Mosses

**JAGDISH LAL**
*Botanical Survey of India*
*Central Circle, Allahabad-211002*

**BISHEN SINGH MAHENDRA PAL SINGH**
23-A, New Connaught Place
Dehra Dun - 248 001 (INDIA)
2005

A Checklist of Indian Mosses




ISBN: 81-211-0439-4

Printed by Gajendra Singh Gahlot at Shiva Offset Press, and composed by Doon Phototype Printers, 14 Old Connaught Place, Dehra Dun, India for Bishen Singh Mahendra Pal Singh, Dehra Dun, India.

*Dedicated*

*to*

*the past, present and future*

***Bryologists***

*of*

*the country*

# PREFACE

Mosses constitute an important component of mountain forest ecosysytem. Varied climatic conditions have bestowed India with rich moss flora, yet there is no published checklist for the country. In the recent years various countries like Japan, Indo China, Europe, China etc. have published their moss checklist but 'India is probably the largest area of land in the world without a published checklist' (O'Shea, J. Hattori Bot. Lab. No. 93 : 293-304, 2003). Bruehl's (1931) "Census of Indian Mosses" is difficult to match to current geographical boundarics and is alomost out of date. Chopra (1975) covered a larger area including the non-Indian Himalayas and mentioned many species as "likely to occur". Gangulee (1969- 80) covered only eastern region.

To fill this lacunae an attempt has been made by the author to compile the moss checklist which is literature based. In moss flora of India *ca* 1576 moss taxa (including varieties, subspecies and forma) are represented which also includes 70 taxa which are confined to nearby eastern Nepal but included by Gangulee (1969-80). If strict geographical configurations are adhered to only 1506 taxa of mosses distributed under 338 genera and 57 families occur in India.

**Jagdish Lal**

# A CHECKLIST OF INDIAN MOSSES

Linnaeus (1753) in his "Species Plantarum" included not only the higher plants but also the mosses. This work is considered as the starting point for the principle of priority for *Sphagnum*, but for other mosses Hedwig's (1801) "Species Muscorum" is the starting point. W. Mitten (1859) published "Musci Indiae Orientalis" which was the most important work since it included all the Indian species known at that time and some from the adjoining countries. He recorded nearly 775 species belonging to 85 genera and 18 families. Brotherus (1924-25) enumerated all genera of mosses of the world known till then. He enumerated 670 genera distributed over 83 families and 16 orders. P. Bruehl (1931) compiled a census of Indian mosses, which is primarily based on Brotherus' work, including genera and species from some of the adjoining countries. He listed 2, 417 species belonging to 371 genera. Gangulee (1969-80) published "Mosses of Eastern India and Adjacent Regions" in eight Fascicles which included 990 species.

Chopra (1975) dealt with nearly 2,000 species belonging to 329 genera under 56 families. He listed many species as "likey to occur" and also many *nomen nuda*.

There is no comprehensive account of Indian mosses at present. However, in the recent past checklist of mosses have appeared for many countries namely Japan, North America, Europe and Azores and Indochina and this prompted the preparation of checklist of Indian mosses to ascertain their exact number in this region.

The systematic arrangement of various taxa (subclasses, orders, suborders, families and genera) follow the sequence given by Vitt (1984). However, some innovations proposed by Buck and Vitt (1986) for pleurocarpous mosses have also been incorporated.

In this checklist all names of genera and species of mosses reported from India are listed in alphabetical order. All scientific names are published in italics; the valid epithets are printed in bold faced italics. Each valid epithert and basionym if any is followed by full citation. Distribution of each species is given in different Bryogeographical units of Indian flora. The present list includes 1576 taxa distributed over 338 genera belonging to 57 families.

## ACKNOWLEDGEMENTS

I am thankful to the Director, Botanical Survery of India, Kolkata for facilities. I am highly indebted to my reverend teachers Prof. N.S. Parihar and Prof. J.P. Srivastava, the renowned bryologists, who made me available their entire personal bryo-literature for this purpose.

**Jagdish Lal**

# CONTENTS

# SYSTEMATIC ARRANGEMENT OF GENERA

## Class - Bryopsida

S.C Sphagnidae
O Sphagnales
F. **Sphagnaceae**
G. *Sphagnum* (22)

S.C. Andreaeidae
O. Andreaeales
F. **Andreaeaceae**
G. *Andreaea* (6)

S.C Takakidae
O. Takakiales
S.O. Takakiineae
F. **Takakiaceae**
G. *Takakia* (2)

S.C Bryidae
O Polytrichales
F. **Polytrichaceae**
G. *Atrichum* (6)
*Lyellia* (2)
*Oligotrichum* (4)
*Pogonatum* (35)
*Polytrichastrum* (4)
*Polytrichum* (5)

O. Bryales
S.O. Archidiineae
F. **Archidiaceae**
G. *Archidium* (4)

S.O. Funariineae
F. **Funariaceae**
G. *Entosthodon* (11)
*Funaria* (9)
*Physcomitrium* (8)

F. **Ephemeraceae**
G. *Micromitrium* (1)

S.O. Splachnineae
F. **Splachnaceae**
G. *Gymnostomiella* (1)
*Splachnobryum* (6)
*Tayloria* (7)
*Tetraplodon* (1)
*Voitia* (1)

S.O. Orthotrichineae
F. **Orthotrichaceae**
G. *Desmotheca* (1)
*Drummondia* (3)
*Groutiella* (1)
*Macrocoma* (1)
*Macromitrium* (11)
*Octogonella* (1)
*Orthotrichum* (19)
*Schlotheimia* (1)
*Trigonodictyon* (1)
*Ulota* (2)
*Zygodon* (10)

F. **Erpodiaceae**
G. *Aulacopilum* (3)
*Erpodium* (1)
*Solmsiella* (2)

F. **Rhachitheciaceae**
G. *Rhachithecium* (1)

S.O. Bryineae
F. **Bryaceae**
G. *Anomobryum* (12)
*Brachymenium* (14)

*Bryum* (37)
*Epipterygium* (1)
*Leptobryum* (1)
*Mielichhobryum* (1)
*Mielichhoferia* (9)
*Mniobryum* (3)
*Plagiobryum* (2)
*Pohlia* (13)
*Rhodobryum* (5)

F. **Mniaceae**
G. *Mnium* (27)
*Orthomniopsis* (1)
*Orthomnium* (1)
*Plagiomnium* (1)

F. **Rhizogoniaceae**
G. *Rhizogonium* (1)

F. **Timmiaceae**
G. *Timmia* (3)

F. **Bartramiaceae**
**G.** *Anacolia* (2)
*Bartramia* (9)
*Bartramidula* (4)
*Breutelia* (3)
*Conostomum* (1)
*Fleischerobryum* (2)
*Philonotis* (26)
*Plagiopus* (1)

F. **Aulacomniaceae**
G. *Aulacomnium* (3)

F. **Meesiaceae**
G. *Amblyodon* (1)
*Meesia* (2)

S.O. Hypnineae
F. **Thuidiaceae**
G. *Abietinella* (2)
*Actinothuidium* (2)
*Claopodium* (5)
*Haplocladium* (6)
*Myurella* (2)
*Pelekium* (2)
*Thuidium* (24)
*Trachyphyllum* (4)

F. **Racopilaceae**
G. *Racopilum* (3)

F. **Myriniaceae**
G. *Schwetschkeopsis* (3)

F. **Lembophyllaceae**
G. *Dixonia* (1)
*Isothecium* (2)

F. **Amblystegiaceae**
G. *Amblystegium* (7)
*Calliergon* (3)
*Calliergonella* (1)
*Campyliadelphus* (2)
*Campylium* (5)
*Cratoneuron* (5)
*Drepanocladus* (4)
*Hygroamblystegium* (3)
*Hygrohypnum* (4)
*Leptodictyum* (1)
*Ortholimnobium* (1)
*Sciaromium* (1)

F. **Brachytheciaceae**
G. *Brachythecium* (36)
*Bryhnia* (3)
*Camptothecium* (1)
*Cirriphyllum* (3)
*Eurhynchium* (10)
*Homalothecium* (6)
*Palamocladium* (2)
*Rhynchostegiella* (11)

*Rhynchostegium* (11)
*Rozea* (1)

F. **Plagiotheciaceae**
G. *Juratzkaea* (1)
*Plagiothecium* (12)
*Stereophyllum* (11)

F. **Hypnaceae**
G. *Bryosedgwickia* (3)
*Ctenidium* (1)
*Ectropothecium* (12)
*Gollania* (3)
*Homomallium* (2)
*Hypnum* (10)
*Isopterygiopsis* (1)
*Isopterygium* (21)
*Leiodontium* (2)
*Leucomium* (2)
*Mittenothamnium* (1)
*Nanothecium* (1)
*Orthothecium* (1)
*Platydictya* (3)
*Platygyrium* (3)
*Pseudostereodon* (1)
*Ptilium* (1)
*Pylaisia* (4)
*Taxiphyllum* (3)
*Vesicularia* (10)

F. **Hylocomiaceae**
G. *Cyathothecium* (1)
*Hylocomium* (3)
*Leptohymenium* (1)
*Macrothamniella* (1)
*Macrothamnium* (5)
*Neodolichomitra* (1)
*Orontobryum* (2)
*Pleurozium* (1)
*Rhytidiadelphus* (1)
*Stenotheciopsis* (1)

F. **Rhytidiaceae**
G. *Rhytidium* (1)

F. **Entodontaceae**
G. *Campylodontium* (2)
*Entodon* (17)
*Erythrodontium* (1)
*Orthotheciadelphus* (1)
*Retidens* (1)

F. **Sematophyllaceae**
G. *Acanthorrhynchium* (1)
*Acroporium* (1)
*Aptychella* (7)
*Brotherella* (12)
*Chinostomum* (1)
*Clastobryopsis* (2)
*Clastobryum* (3)
*Foreauella* (1)
*Gammiella* (1)
*Glossadelphus* (8)
*Hageniella* (3)
*Heterophyllium* (3)
*Meiothecium* (3)
*Pylaisiadelpha* (1)
*Pylaisiopsis* (1)
*Rhaphidorrhynchium* (1)
*Rhaphidostichum* (5)
*Sematophyllum* (8)
*Struckia* (2)
*Taxithelium* (6)
*Trichosteleum* (8)
*Trismegistia* (1)
*Trolliella* (1)
*Warburgiella* (4)
*Wijkia* (8)

S.O. Leucodontineae
F. **Leskeaceae**
G. *Anomodon* (8)

*Habrodon* (1)
*Haplohymenium* (1)
*Herpetineuron* (1)
*Leptopterigynandrum* (4)
*Lescuraea* (6)
*Leskea* (3)
*Leskeella* (2)
*Lindbergia* (3)
*Okamuraea* (1)
*Pseudoleskea* (3)
*Pseudoleskeella* (1)
*Pseudoleskeopsis* (4)
*Schwetschkea* (2)

F. **Pterigynandraceae**
G. *Pterigynandrum* (2)

F. **Fabroniaceae**
G. *Anacamptodon* (1)
*Fabronia* (10)
*Levierella* (1)

F. **Cryphaeaceae**
G. *Hydrocryphaea* (1)
*Pilotrichopsis* (2)
*Schoenobryum* (2)
*Sphaerotheciella* (1)

F. **Leucodontaceae**
G. *Forsstroemia* (3)
*Leucodon* (2)

F. **Regmatodontaceae**
G. *Regmatodon* (2)

F. **Hedwigiaceae**
G. *Braunia* (3)
*Hedwigia* (1)
*Hedwigidium* (1)
*Kashyapia* (1)

F. **Trachypodaceae**
G. *Diaphanodon* (2)
*Duthiella* (4)
*Pseudospiridentopsis* (1)
*Trachypodopsis* (2)
*Trachypus* (3)

F. **Pterobryaceae**
G. *Calyptothecium* (9)
*Endotrichella* (1)
*Garovaglia* (1)
*Horikawaea* (1)
*Jaegerina* (1)
*Meteoriella* (1)
*Myurium* (3)
*Osterwaldiella* (1)
*Penzigiella* (1)
*Pterobryopsis* (9)
*Symphysodontella* (5)
*Trachyloma* (1)

F. **Meteoriaceae**
G. *Aerobryidium* (1)
*Aerobryopsis* (4)
*Aerobryum* (2)
*Barbella* (15)
*Chrysocladium* (6)
*Floribundaria* (10)
*Meteoriopsis* (5)
*Meteorium* (4)
*Neobarbella* (1)
*Papillaria* (5)
*Pseudobarbella* (2)

F. **Neckeraceae**
G. *Cryptoleptodon* (3)
*Handeliobryum* (3)
*Himantocladium* (6)
*Homalia* (2)

*Homaliadelphus* (1)
*Homaliodendron* (14)
*Neckera* (9)
*Neckeropsis* (9)
*Pinnatella* (10)
*Porotrichum* (1)
*Thamnobryum* (7)

S.O. Hookeriineae
F. **Daltoniaceae**
G. *Actinodontium* (1)
*Daltonia* (13)

F. **Symphyodontaceae**
G. *Symphyodon* (8)

F. **Hookeriaceae**
G. *Callicostella* (1)
*Chaetomitriopsis* (1)
*Chaetomitrium* (2)
*Distichophyllum* (12)
*Eriopus* (2)
*Hookeria* (1)
*Hookeriopsis* (2)
*Lepidopilidium* (1)

F. **Hypopterygiaceae**
G. *Cyathophorella* (6)
*Dendrocyathophorum* (2)
*Hypopterygium* (3)

S.O. Buxbaumiineae
F. **Buxbaumiaceae**
G. *Buxbaumia* (1)
*Diphyscium* (3)
*Theriotia* (2)

S.O. Encalyptineae
F. **Encalyptaceae**
G. *Encalypta* (6)

S.O. Pottiineae
F. **Pottiaceae**
G. *Aloina* (1)
*Anoectangium* (8)
*Astomum* (2)
*Barbula* (22)
*Bellibarbula* (2)
*Bryoerythrophyllum* (9)
*Cinclidotus* (1)
*Desmatodon* (4)
*Didymodon* (15)
*Eucladium* (1)
*Ganguleea* (1)
*Gymnostomum* (2)
*Hydrogonium* (12)
*Hymenostomum* (4)
*Hymenostyliella* (1)
*Hymenostylium* (8)
*Hyophila* (6)
*Hyophilopsis* (1)
*Leptodontium* (3)
*Molendoa* (3)
*Oxystegus* (8)
*Pleurochaete* (1)
*Pottia* (3)
*Prionidium* (1)
*Pseudocrossidium* (1)
*Pseudosymblepharis* (4)
*Reimersia* (1)
*Rhamphidium* (4)
*Scopelophila* (2)
*Semibarbula* (2)
*Stegonia* (1)
*Syntrichia* (1)
*Timmiella* (3)
*Tortella* (6)
*Tortula* (15)
*Trachycarpidium* (1)
*Trichostomum* (9)
*Tuerckheimia* (1)
*Weissia* (5)

F. **Calymperaceae**
G. *Calymperes* (15)
*Calymperopsis* (1)
*Exodictyon* (1)
*Leucophanes* (4)
*Mitthyridium* (6)
*Octoblepharum* (1)
*Syrrhopodon* (11)

S.O. Dicranineae
F. ***Dicranaceae***
G. *Anisothecium* (1)
*Aongstroemia* (2)
*Aongstroemiopsis* (1)
*Atractylocarpus* (2)
*Brothera* (3)
*Campylopodiella* (2)
*Campylopodium* (3)
*Campylopus* (20)
*Cynodontium* (2)
*Dicranella* (8)
*Dicranodontium* (13)
*Dicranoloma* (5)
*Dicranoweisia* (5)
*Dicranum* (13)
*Holomitrium* (1)
*Leucoloma* (10)
*Microcampylopus* (1)
*Microdus* (2)
*Mitrobryum* (1)
*Onchophorus* (3)
*Oreas* (1)
*Oreoweisia* (2)
*Orthodicranum* (1)
*Paraleucobryum* (2)
*Symblepharis* (2)
*Trematodon* (12)
*Wilsoniella* (1)

F. **Leucobryaceae**
G. *Leucobryum* (12)
*Ochrobryum* (2)
*Thiemea* (1)

F. **Rhabdoweisiaceae**
G. *Amphidium* (2)
*Rhabdoweisia* (2)

F. **Ditrichaceae**
G. *Ceratodon* (2)
*Distichium* (2)
*Ditrichopsis* (1)
*Ditrichum* (12)
*Garckea* (2)
*Pleuridiella* (1)
*Pleuridium* (3)
*Saelania* (1)

S.O. Fissidentineae
F. **Fissidentaceae**
G. *Fissidens* (67)

S.O. Seligeriineae
F. **Seligeriaceae**
G. *Blindia* (4)

S.O. Grimmiineae
F. **Ptychomitriaceae**
G. *Ptychomitrium* (3)

F. **Grimmiaceae**
G. *Coscinodon* (1)
*Grimmia* (20)
*Racomitrium* (9)
*Schistidium* (2)

---

S.C. : Subclass
O. : Order
S.O. : Suborder
( ) : No. of taxa in brackets
F. : Family
G. : Genus

# FAMILIES OF INDIAN MOSSES

Amblystegiaceae G. Roth, Hedwigia 38(6): 1899. T: *Amblystegium* B.S.G.

Andreaeaceae Dum., Anal. Fam. Pl. 68. 1829. T: *Andreaea* Hedw.

Archidiaceae Schimp., Coroll. Bryol. Eur. 5. 1855-1856. T: *Archidium* Brid.

Aulacomniaceae Schimp., Syn. Musc. Eur. 411. 1860. T: *Aulacomnium* Schwaegr.

Bartramiaceae Schwaegr. in Willd., Sp. Pl. ed. 4. 5(2): 90. 1830. T: *Bartramia* Hedw.

Brachytheciaceae Schimp., Syn. Musc. Eur., ed. 2. xcv sic (cxv), 637. 1876. T: *Brachythecium* B.S.G.

Bryaceae Schwaeger. in Willd., Sp. Pl. ed. 4. 5(2): 47. 1830. T: *Bryum* Hedw.

Buxbaumiaceae Schwaegr. in Willd., Sp. Pl. ed. 4. 5(2): 23. 1830. T: *Buxbaumia* Hedw.

Calymperaceae Kindb., Gen. Eur. Northamer. Bry. 11. 1897. T: *Calymperes* Sw. in Web.

Cryphaeaceae Schimp., Coroll. Bryol. Eur. 97. 1855-1856. T: *Cryphaea* Mohr in Web.

Daltoniaceae Schimp., Syn. Musc. Eur. 478. 1860. T : *Daltonia* Hook. & Tayl.

Dicranaceae Schimp., Coroll. Bryol. Eur. 11. 1855-1856. T: *Dicranum* Hedw.

Ditrichaceae Limpr. in Rabenh., Kryptogamen- Fl. ed. 2. 4 : 482. 1887. T: *Ditrichum* Hampe.

Encalyptaceae Schimp., Coroll. Bryol. Eur. 38. 1855-1856. T: *Encalypta* Hedw.

Entodontaceae Kindb., Gen. Eur. Northamer. Bry. 7. 1897. T: *Entodon* C. Muell.

Ephemeraceae Schimp., Coroll. Bryol. Eur. 3. 1855 1856. T: *Ephemerum* Hampe

Erpodiaceae Broth., Nat. Pfl.I(3): 706. 1905. T: *Erpodium* (Brid.) Brid.

Fabroniaceae Schimp., Coroll. Bryol. Eur. 102. 1855 1856. T: *Fabronia* Raddi

Fissidentaceae Schimp., Coroll. Bryol. Eur. 20. 1855 1856. T: *Fissidens* Hedw.

Funariaceae Schwaegr. in Willd., Sp. Pl. ed. 4. 5(2): 43. 1830. T: *Funaria* Hedw.

Grimmiaceae Arnott, Disp. Meth. Mousses 19. 1825. T: *Grimmia* Hedw.

Hedwigiaceae Schimp., Coroll. Bryol. Eur. 52. 1855 (1856). T: *Hedwigia* P. Beauv.

Hookeriaceae Schimp., Coroll. Bryol. Eur. 101. 1855 (1856). T: *Hookeria* J. E. Sm.

Hylocomiaceae (Broth.) Fleisch., Nova Guinea 12: 125. 1914. T: *Hylocomium* B.S.G.

Hypnaceae Schimp., Coroll. Bryol. Eur. 113. 1855 (1856). T. *Hypnum* Hedw.

Hypopterygiaceae Mitt., Jour. Proc. Linn. Soc. Suppl. Bot. 1: 147. 1859. T: *Hypopterygium* Brid.

Lembophyllaceae Broth., Nat. Pfl. 1(3): 863. 1906. T: *Lembophyllum* Lindb.

Leskeaceae Schimp., Coroll. Bryol. Eur. 109. 1855 (1856). T: *Leskea* Hedw.

Leucobryaceae Schimp., Coroll. Bryol. Eur. 19. 1855 (1856). T: *Leucobryum* Hampe

Leucodontaceae Schimp., Coroll. Bryol. Eur. 108. 1855 (1856). T: *Leucodon* Schwaegr.

Meesiaceae Schimp., Coroll. Bryol. Eur. 82. 1855 1856. T: *Meesia* Hedw.

Meteoriaceae Kindb., Gen. Eur. Northamer. Bry. 7. 1897. T: *Meteorium* (Brid.) Doz. et Molk.

Mniaceae Schwaegr. in Willd., Sp. Pl. ed. 4. 5(2): 25. 1830. T: *Mnium* Hedw.

Myriniaceae Schimp., Syn. Musc. Eur. ed. 2, c XI, 589. 1876.

Neckeraceae Schimp., Coroll. Bryol. Eur. 99. 1855-1856. T: *Neckera* Hedw.

Orthotrichaceae Arnott, Disp. Meth. Mousses 13. 1825. T: *Orthotrichum* Hedw.

Plagiotheciaceae (Broth.) Fleisch., Nova Guinea 8 : 748. 1912. T: *Plagiothecium* B.S.G.

Polytrichaceae Schwaegr. in Willd., Sp. Pl. ed. 4. 5(2): 1. 1830. T: *Polytrichum* Hedw.

Pottiaceae Schimp., Coroll. Bryol. Eur. 24. 1855 1856. T: *Pottia* (Reichenb.) Furnr.

Pterigynandraceae Schimp., Syn. Musc. Eur. ed.2, c XIII, 618. 1876.

Pterobryaceae Kindb., Sp. Eur. Northamer. Bry. 15. 1896. T: *Pterobryon* Hornsch.

Ptychomitriaceae Schimp., Syn. Musc. Eur. 241. 1860. T: *Ptychomitrium* Furnr.

Racopilaceae Kindb., Bot. Centralbl. 76: 85. 1898. T: *Racopilum* P. Beauv.

Regmatodontaceae Broth., Nat. Pfl. I(3): 1125. 1908. T: *Regmatodon* Brid.

Rhabdoweisiaceae Limpr. *in* Rabenh. Kryptogamen Fl., ed. 2. 4: 271. 1886. T: *Rhabdoweisia* B.S.G.

Rhachitheciaceae Robins., Bryologist 67: 448. 1964. T: *Rhachithecium* Broth. *ex* Le Jol.

Rhizogoniaceae Broth., Nat. Pfl. 1(3): 614. 1904. T: *Rhizogonium* Brid.

Rhytidiaceae Broth., Nat. Pfl. ed. 2. 11: 475. 1925. T: *Rhytidium* (Sull.) Kindb.

Seligeriaceae Schimp., Coroll. Bryol. Eur. 22. 1855. 1856. T: *Seligeria* B.S.G.

Sematophyllaceae Broth., Nat. Pfl. 1(3): 1098. 1908. T: *Sematophyllum* Mitt.

Sphagnaceae Dum., Anal. Fam. Pl. 68. 1829. T: *Sphagnum* L.

Splachnaceae Grev. & Arnott, Mem. Wernerian Soc. 5: 442. 1824. T: *Splachnum* Hedw.

Symphyodontaceae Fleisch., Musci Fl. Buitenzong 4: 1110. 1923. T: *Symphyodon* Mont.

Takakiaceae Hatt. *et* Inoue, J. Hattori Bot. Lab. 19 : 137. 1958. T: *Takakia* Hatt. *et* Inoue.

Thuidiaceae Schimp., Syn. Musc. Eur. 493. 1860. T: *Thuidium* B.S.G.

Timmiaceae Schimp., Coroll. Bryol. Eur. 87. 1855. 1856. T: *Timmia* Hedw.

Trachypodaceae Fleisch., Hedwigia 45: 63. 1906. 1905. T: *Trachypus* Reinw. et Hornsch.

# ABBREVIATIONS

In the alphabetical list, against each species is indicated its distribution in different bryogeographical units. The list of abbreviations for bryogeographical units is as follows :

W.H. – The Western Himalayan territory extending from the western boundary of Nepal to Kashmir.

E.H. – The Eastern Himalayan territory comprising eastern part of mountains of Indian territory which separate India proper from Burma and China.

P. – Punjab Plain.

R. – Rajasthan.

G.P. – Gangetic Plain.

S.I. – South Indian Zone (South of Vindhyas and Rajmahal hills and extending South to Cape Comorin.

I.O. – India Orientalis (where the locality is not definitely known but reported from somewhere in the old Indian Empire, including the neighbouring region of Burma, Pakistan etc. by some early workers, e.g. Mitten.

C.I. – Central Indian Zone.

# BRYOGEOGRAPHICAL UNITS OF INDIAN FLORA

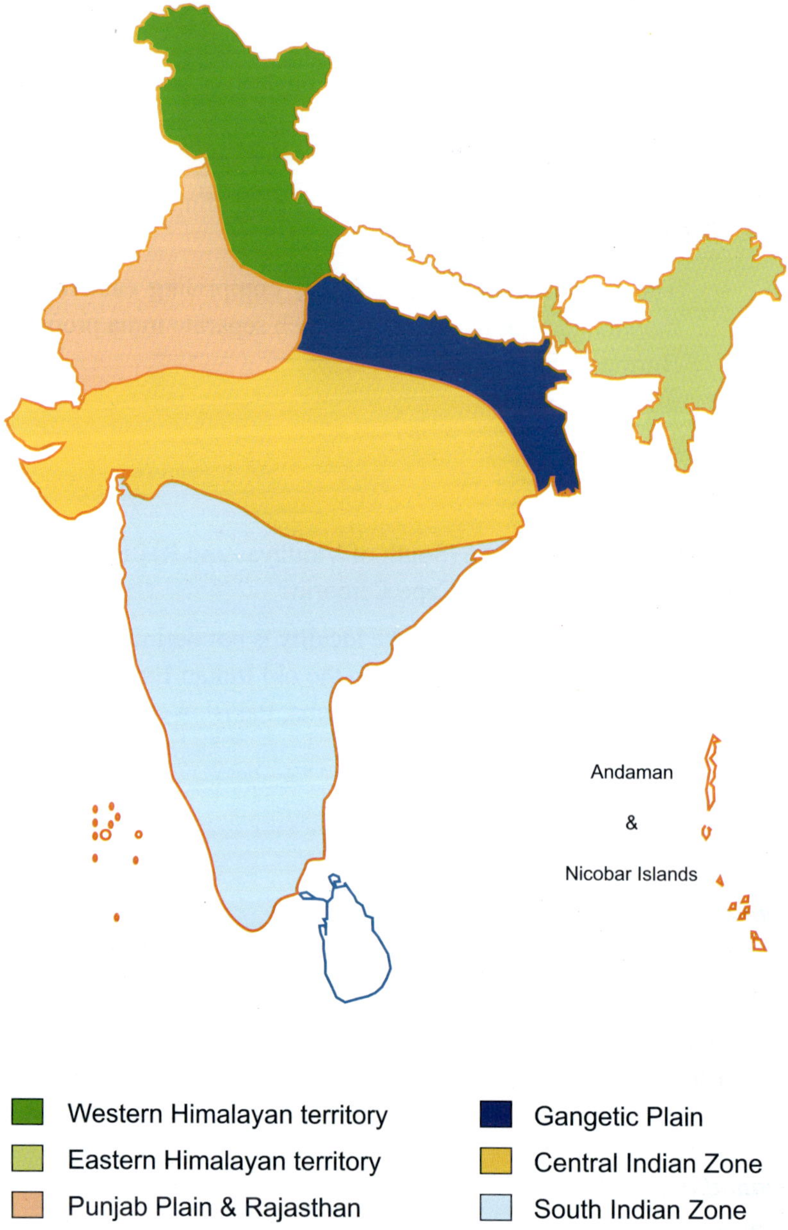


Map not to scale

# ALPHABETICAL ARRANGEMENT OF GENERA AND SPECIES OF INDIAN MOSSES

***Abietinella*** C. Muell., Nuov. Giorn. Bot. Ital. n. ser. 3: 115. 1896.
(Thuidiaceae)

Type : *A. giraldii* C. Muell.

– ***abietina*** (Hedw.) Fleisch., Musci Fl. Buitenzorg 4: 1497. 1923. *Hypnum abietinum* L. *ex* Hedw., Spec. Musc. 353. 1801.
........ W.H.; E.H.; Nepal

– ***brandisii*** (C. Muell.) Vohra, Bull. Bot. Surv. India 21(1-4): 171. 1979. *Hypnum brandisii* C. Muell., Flora 61: 86. 1878.
........ W.H. Endemic

***Acanthorrhynchium*** Fleisch., Musci Fl. Buitenzorg 4: 1331. 1923.
(Sematophyllaceae)

Type : *A. papillatum* (Harv.) Fleisch.

– ***papillatum*** (Harv.) Fleisch., Musci Fl. Buitenzorg 4: 1331. 216. 1923. *Hypnum papillatum* Harv. in Hook., Icon. Pl. Rar. 1: 23. f.8. 1836. ........ E.H.; Nicobar Is.; Nepal

***Acroporium*** Mitt., J. Linn. Soc. Bot. 10: 182. 1868. (Sematophyllaceae)

– ***baviense*** (Besch.) Broth., Nat. Pfl. ed. 2, 11: 437. 1925. *Sematophyllum baviense* Besch., J. de Bot. 4: 205. 1890.
........ E.H.

***Actinodontium*** Schwaegr., Spec. Musc. Suppl. 2(2): 75. 1826.
(Daltoniaceae)

Type : *A. adscendens* Schwaegr.

– ***rhaphidostegium*** (C.Muell.) Bosch *et* Lac., Bryol. Jav. 2: 37. 160. 1862. *Hookeria rhaphidostega* C. Muell., Syn. 2: 677. 1851.
........ E.H.; S.I.

***Actinothuidium*** (Besch.) Broth., Nat. Pfl. 1(3): 1019. 1908. (Thuidiaceae)

Type : *A. hookeri* (Mitt.) Broth.

– ***hookeri*** (Mitt.) Broth., Nat. Pfl. 1(3): 1020. 1908. *Leskea hookeri* Mitt., Linn. Soc. Bot. Suppl. 1: 132. 1859. ........ E.H.; Nepal

– ***sikkimense*** Warnst., Hedwigia 57: 112. 46. 1915.
......... E.H. Endemic

***Aerobryidium*** Fleisch. *in* Broth., Nat. Pfl. 1(3): 820. 1906. (Meteoriaceae)

L. Type : *A. filamentosum* (Hook.) Fleisch.

– ***filamentosum*** (Hook.) Fleisch. *in* Broth., Nat. Pfl. 1(3): 821. 1906. *Neckera filamentosa* Hook., Musci Exot. 2: 158. 1819.
........ W.H.; E.H.; S.I.; Nepal

***Aerobryopsis*** Fleisch., Hedwigia 44: 304. 1905. (Meteoriaceae)

L.Type : *A. longissima* (Doz. et Molk.) Fleisch.

– ***denticulata*** Dix., J. Bombay Nat. Hist. Soc. 39(4): 783. 1937.
........E.H.

– ***longissima*** (Doz. *et* Molk.) Fleisch., Hedwigia 44: 305. 1905. *Neckera longissima* Doz. et Molk., Ann. Sc. Nat. Bot. ser. 3, 2: 313. 1844. ........ W.H.; E.H.; S.I.

– ***membranacea*** (Mitt.) Broth., Nat. Pfl. 1(3): 819. 1906. *Meteorium membranaceum* Mitt., J. Linn. Soc. Bot. Suppl. 1: 88. 1859.
........ E.H. Endemic

– ***wallichii*** (Brid). Fleisch., Musci Fl. Buitenzorg 3:789. 1908. *Hypnum wallichii* Brid., Bryol. Univ. 2 : 416. 1827. ........ Nepal

***Aerobryum*** Doz. *et* Molk., Ned. Kruidk. Arch. 2(4): 279. 1851.
(Meteoriaceae)

Type : *A.speciosum* Doz. *et* Molk.

– ***speciosum*** Doz. *et* Molk., Ned. Kruidk. Arch. 2(4): 279. 1851. *Meteorium speciosum* (Doz. *et* Molk.) Mitt., J. Linn. Soc. Bot. Suppl. 1: 87. 1859. ........ E.H.

– ***willisii*** Fleisch., Musci Fl. Buitenzorg 3: 846. 1908. ........ S.I.

***Aloina*** (C. Muell.) Kindb., Bih. K. Svensk. Vet. Ak. Handl. 6(19): 22. 1882. (Pottiaceae)

L. Type : *A. aloides* (Schultz) Kindb.

– ***rigida*** (Hedw.) Limpr., Laubm. Deutschl. 4(1): 637. 1888. *Barbula rigida* Hedw., Spec. Musc. 115. 1801. ........ W.H.

***Amblyodon*** B.S.G., Bryol. Eur. 4: 3. 307. 1841. (Meesiaceae)

Type : *A. dealbatus* (Hedw.) B.S.G.

– ***dealbatus*** (Hedw.) B.S.G., Bryol. Eur. 4: 7. 307. 1841. *Meesia dealbata* Sw. ex Hedw., Spec. Musc. 174. 1801. ........ W.H.

***Amblystegium*** B.S.G., Bryol. Eur. 6: 45. 1853. (Amblystegiaceae)

L. Type: *A. serpens* (Hedw.) B.S.G.

– ***juratzkanum*** Schimp., Syn. 693. 1860. ........ W.H.

– ***saxatile*** Schimp., Syn. 595. 1860. ........ W.H.

– ***serpens*** (Hedw.) B.S.G., Bryol. Eur. 6: 53. 564. 1853. *Hypnum serpens* L. ex Hedw., Spec. Musc. 268. 1801. ........ W.H.

var. ***saxicola*** (A. Hans.) C. Jens., Danm. Moss. 2: 58. 1923. *Amblystegium saxicola* A. Hans., Bot. Tidskr. 25: 405. 6. 1903. ........ W.H.

– ***sparsile*** (Mitt.) Jaeg., Ber. S. Gall. Naturw. Ges. 1877-78: 290. 1880. *Hypnum sparsile* Mitt., J. Linn. Soc. Bot. Suppl. 1: 82. 1859. ........ E.H.

– ***tibetanum*** (Mitt.) Par., Ind. Bryol. 23. 1894. *Hypnum tibetanum* Mitt., J.Linn. Soc. Bot. Suppl. 1: 83. 1859. ........ W.H.

– ***varium*** (Hedw.) Lindb., Musci Scand. 32. 1879. *Leskea varia* Hedw., Spec. Musc. 216. 1801. ........ W.H.

***Amphidium*** Schimp., Coroll. 39. 1856. (Rhabdoweisiaceae)

L. Type : *A. lapponicum* (Hedw.) Schimp.

– ***lapponicum*** (Hedw.) Schimp., Coroll. 39. 1856. *Anictangium lapponicum* Hedw., Spec. Musc. 40. 1801. ........ W.H.

– ***mougeotii*** (B.S.G.) Schimp., Coroll. 40. 1856. *Zygodon mougeotii* B.S.G., Bryol. Eur. 3: 39. 206. 1838. ........ Nepal

***Anacamptodon*** Brid., Mant. Musc. 136.2 f.11. 1819. (Fabroniaceae)

Type : *A. splachnoides* (Brid.) Brid.

– ***validinervis*** Dix. *et* P. Vard., Arch. Bot. 1(8-9): 172. 5 f. 4. 1927. ........ S.I.

***Anacolia*** Schimp., Syn. ed. 2: 513. 1876. (Bartramiaceae)

L. Type : *A. webbii* (Mont.) Schimp.

– ***menziesii*** (Turn.) Par., Ind. Bryol. 27. 1894. *Bartramia menziesii* Turn., Ann. Bot. 1:525. 11f. 1. 1805. ........ Nepal

– ***sinensis*** Broth., Wiss. Wien, Math. Nat. Kl. Abt. 1, 133: 570. 1924. ........ E.H.; Nepal

***Andreaea*** Hedw., Spec. Musc. 47. 1801. (Andreaeaceae)

L. Type : *A.rupestris* Hedw.

– ***commutata*** C. Muell., Bot. Zeit. 22: 373. 1864. ........ E.H. Endemic

– ***densifolia*** Mitt., J. Linn. Soc. Bot. Suppl. 1: 7. 1859. ........ E.H. Endemic

– ***indica*** Mitt., J. Linn. Soc. Bot. Suppl. 1: 6. 1859. ........ E.H. Endemic

– ***kashyapii*** Dix. *ex* Vohra & Wadhwa, Bull. Bot. Surv. India 6: 321. 1964. ........ W.H. Endemic

– ***rigida*** Wils. *in* Mitt., J. Linn. Soc. Bot. Suppl. 1: 6. 1859. ........ E.H. Endemic

– ***rupestris*** Hedw., Spec. Musc. 47. 7f. 2 g-p. 1801. ........ W.H.; E.H.; Nepal

***Anisothecium*** Mitt., J. Linn. Soc. Bot. 12: 39. 1869. (Dicranaceae)

– ***molliculum*** (Mitt.) Broth., Nat. Pfl. ed. 2, 10: 177. 1924. *Leptotrichum molliculum* Mitt., Musc. Ind. Or. 11. 1859. ........ W.H.; E.H.

***Anoectangium*** Schwaegr., Spec. Musc. Suppl. 1(1): 33. 1811. (Pottiaceae)

L. Type : *A. compactum* Schwaegr.

– ***bicolor*** Ren. *et* Card., Bull. Soc. R. Bot. Belg. 47(1): 9. 1905. ........ W.H.; E.H.

– ***clarum*** Mitt., J. Linn. Soc. Bot. Suppl. 1: 31. 1859. ........ W.H.; E.H.; Nepal

– ***euchloron*** (Schwaegr.) Mitt., J. Linn. Soc. Bot. 12: 176. 1869. *Gymnostomum euchloron* Schwaegr., Spec. Musc. Suppl. 2(2): 83. 177. 1827. ........ S.I.

– ***kashmiriense*** Aziz & Vohra, Bull. Bot. Surv. India 25: 239. 1983. ........ W.H.

– ***sikkimense*** Aziz & Vohra, Bull. Bot. Surv. India 30: 185. (1988) 1990 ........ E.H.

– ***stracheyanum*** Mitt., J. Linn. Soc. Bot. Suppl. 1: 31. 1859. ........ W.H.; E.H.; S.I.; Nepal

– ***thomsonii*** Mitt., J. Linn. Soc. Bot. Suppl. 1: 31. 1859. ........W.H.; E.H.; S.I.; Nepal

– ***walkeri*** Broth., Rec. Bot. Surv. India 1(12): 311. 1899. ........ S.I.

***Anomobryum*** Schimp., Syn. 382. 1860. (Bryaceae)

Type : *A. julaceum* (Brid.) Schimp.

– ***astorense*** (Broth.) Broth., Nat. Pfl. 1(3): 563. 1903. *Bryum astorense* Broth., Act. Soc. Sc. Fenn. 24(2): 26. 1899. ........ W.H.

– ***auratum*** (Mitt.) Jaeg., Ber. S. Gall. Naturw. Ges. 1873-74: 142. 1875. *Bryum auratum* Mitt., J. Linn. Soc. Bot. Suppl. 1: 67. 1859. ........ W.H.; E.H.; S.I.; Nepal

– ***brachymenioides*** Dix. *et* P. Vard., Ann. Crypt. Exot. 3: 177. 3f.4. 1930. ........ S.I.

– ***cymbifolium*** (Lindb.) Broth., Rec. Bot. Surv. India 1(12): 320. 1899. *Bryum cymbifolium* Lindb., Not. Saellsk. F. Fl. Fenn. 11: 45. 1870. ........ W.H.; E.H.; S.I.

– ***filiforme*** (Dicks). Solms *in* Rabenh., Bryoth. Eur. 25: 331. 1873. *Bryum filiforme* Dicks., Pl. Crypt. Brit. 4: 16. 1801. ........W.H.; E.H.; S.I.; Nepal

var. ***concinnatum*** (Spruc). Loesk., Rev. Bryol. 20:43. 1893. *Bryum concinnatum* Spruc., Musci Pyren. n. 121. 1847. ........ W.H.; E.H.; S.I.; Nepal

– ***gemmigerum*** Broth., Ohilopp. J. Sc. 5(2C): 146. 1910. ........ I.O.

– ***kashmirense*** (Broth.) Broth., Nat. Pfl. 1(3): 563. 1903. *Bryum kashmirense* Broth., Act. Soc. Sc. Fenn. 24(2): 24. 1899. ........ W.H.

– ***latifolium*** Card. *et* P. Vard., Rev. Bryol. 50(2): 19.5. 1923. ........ S.I.

– ***marginatum*** Dix. et Badhw., Rec. Bot. Surv. India 12(2): 173. 1938. ........W.H.

– ***nitidum*** (Mitt.) Jaeg., Ber. S. Gall. Naturw. Ges. 1873-74: 142. 1875. *Bryum nitidum* Mitt., J. Linn. Soc. Bot. Suppl. 1: 67. 1859. ........ W.H.; E.H.

– ***pellucidum*** Dix. et Badhw., Rec. Bot. Surv. India 12 : 173. 1938. ........W. H.

***Anomodon*** Hook. *et* Tayl., Musc. Brit. 79. 3. 1818. (Leskeaceae)

L.Type : *A. viticulosus* (Hedw.) Hook. *et* Tayl.

– ***acutifolius*** Mitt., J. Linn. Soc. Bot. Suppl. 1: 126. 1859. ........ W.H.; E.H.; Nepal

– ***attenuatus*** (Hedw.) Hueb., Musc. Germ. 562. 1833. *Leskea attenuata* Hedw., Spec. Musc. 230. 1801. ........ W. H.

– ***minor*** (Hedw.) Lindb. ssp. ***integerrimus*** (Mitt.) Iwats., Hattori Bot. Lab. 26: 4. 1963. *A. integerrimus* Mitt., J. Linn. Soc. Bot. Suppl. 1: 126. 1859. ........ W.H.; E.H.; Nepal

– ***planatus*** Mitt., J. Linn. Soc. Bot. Suppl. 1: 126. 1859. ........ W.H.; E.H.

– ***rostratus*** (Hedw.) Schimp., Syn. 488. 1860. *Leskea rostrata* Hedw., Spec. Musc. 225. 55 f. 13-18. 1801. ........W.H.

– ***rugelii*** (C.Muell.) Keissl., Ann. Naturh. Hofmus. Wien 15: 214. 1900. *Hypnum rugelii* C.Muell., Syn. 2: 472. 1851. ........ W.H.; E.H.; Nepal

– ***thraustus*** C.Muell., Nuov. Giorn. Bot. Ital. n. ser. 5: 207. 1898. ........ W.H.

– ***viticulosus*** (Hedw.) Hook. *et* Tayl., Musc. Brit. 79. 22. 1818. *Neckera viticulosa* Hedw., Spec. Musc. 209. 1801. ........ W.H.; E.H.; Nepal

***Aongstroemia*** B.S.G., Bryol. Eur. 1: 171. 1846. (Dicranaceae)

Type: *A. longipes* (Somm.) B.S.G.

– ***julacea*** (Hook.) Mitt., J. Linn. Soc. Bot. 12: 27. 1869. *Gymnostomum julaceum* Hook., Musc. Exot. l: tab. 42. 1818. ........ Nepal

– ***orientalis*** Mitt., Trans. Linn. Soc. ser. 2, 3: 154. 1891. ........ W.H.; E.H.

***Aongstroemiopsis*** Fleisch., Musci Fl. Buitenzorg 1:331. 1904. (Dicranaceae)

Type: *A. julacea* (Doz. et Molk.) Fleisch.

– ***julacea*** (Doz. et Molk.) Fleisch., Musci Fl. Buitenzorg 1: 331. 58. 1904. *Pottia julacea* Doz. *et* Molk., Fl. Jungh. 335. 1854. ........ Nepal

***Aptychella*** (Broth.) Herz., Biblioth. Bot. 87:157. 1916. (Sematophyllaceae)

Type: *A. proligera* (Broth.) Herz.

– ***borii*** Dix., J. Bombay Nat. Hist. Soc. 39:790. If. 17. 1937. ........ E.H. Endemic

– ***delicata*** (Fleisch.) Fleisch., Musci Fl. Buitenzorg 4: 1671. 1923. *Clastobryopsis delicata* Fleisch., Ibid. 1180. 1923. ........ E.H.

– ***muelleri*** Dix., Ann. Bryol. 6: 32. 1933. ........ E.H.; Nepal

– ***planula*** (Mitt.) Fleisch., Musci Fl. Buitenzorg 4. 1671. 1923. *Stereodon planulus* Mitt., J. Linn. Soc. Bot. Suppl. 1: 111. 1859. ........ E.H.; Nepal

– ***serrulata*** (Card. et P. Vard.) Broth., Nat. Pfl. ed. 2, 11: 535. 1925. *Clastobryum serrulatum* Card. *et* P. Vard., Rev. Bryol. 50: 75. 1923. ........ S.I.

– ***subdelicata*** Broth., Symb. Sin. 4: 117. 1929. ........ Nepal

– ***tenuiramea*** (Mitt.) Tix., Rev. Bryol. Lich. 43: 423. 1977. *Stereodon tenuirameus* Mitt., Musci Ind. Or. 94. 1859 ........ W.H.; E.H.; S.I.; Nepal

***Archidium*** Brid., Bryol. Univ. 1: 747. 1826. (Archidiaceae)

Type: *A. phascoides* Brid.

– ***birmannicum*** Mitt. ex Dix., J. Indian Bot. 2: 175.1. 1921. ........ G.P.; S.I.; Nepal

– ***indicum*** C. Muell., Flora 71:8. 1888. ........ W.H.; C.I.

– ***microthecium*** Dix. et P. Vard., Ann. Crypt. Exot. 1: 37.1. 1928. ........ S.I.

– ***octosporum*** Dix. et P. Vard., Ann. Crypt. Exot. 1: 37.1. 1928. ........ S.I.

***Astomum*** Hamp., Flora 20: 285. 1837. (Pottiaceae)

L. Type : *A. crispum* (Hedw.) Hamp.

- ***crispum*** (Hedw.) Hamp., Flora 20: 285. 1837. *Phascum crispum* Hedw., Spec. Musc. 21.1801. ........ W. H.
- ***minutum*** Dix. *et* P.Vard., Arch. Bot. 1(8-9): 165.1.1927. ........ S.I.

***Atractylocarpus*** Mitt., J. Linn. Soc. Bot. 12: 71. 1869. (Dicranaceae)

Type: *A. mexicanus* Mitt.

- ***erectifolius*** Mitt. *ex* Dix., Not. R.Bot. Gard. Edinburgh 19: 281. 1938. ........ E.H. Endemic
- ***sinensis*** (Broth.) Herz., Ann. Bryol. 12: 87. 1939. *Metzlerella sinensis* Broth., Symb. Sin. 4:19. 1929. ....... E.H.

***Atrichum*** P. Beauv., Mag. Encycl. 5: 329. 1804. (Polytrichaceae)

L. Type: *A. undulatum* (Hedw.) P. Beauv.

- ***aculeatum*** Card. et P. Vard., Rev. Bryol. 50: 25. 14. 1923. ........ W.H.; S.I.
- ***flavisetum*** Mitt., J. Linn. Soc. Bot. Suppl. 1: 150. 1859. ........ W.H.; E.H.; Nepal
- ***obtusulum*** (C.Muell.) Jaeg., Ber. S. Gall. Naturw. Ges. 1877-78: 453. 1880. *Catharinea obtusula* C. Muell., Flora 61: 81. 1878. ........ W.H.; E.H.; Nepal
- ***pallidum*** Ren. *et* Card., Bull. Soc. R. Bot. Belg. 34(2): 63. 1896. .......W.H.; E.H.; Nepal
- ***undulatum*** (Hedw.) P.Beauv., Prodr. 42. 1805. *Polytrichum undulatum* Hedw., Spec. Musc. 198. 1801. ........ W.H.; E.H.; Nepal

  var. ***subserratum*** (Hook.) Par., Ind. Bryol. 55. 1894. *Polytrichum undulatum* Hedw. var. *subserratum* Hook., J. Bot. 2: 3. 1840. ........ W.H.; E.H.; Nepal

***Aulacomnium*** Schwaegr., Spec. Musc. Suppl. 3(1): 215. 1827. (Aulacomniaceae)

L. Type: *A. androgynum* (Hedw.) Schwaegr.

- ***androgynum*** (Hedw.) Schwaegr., Spec. Musc. Suppl. 3(1): 215. 1827. *Bryum androgynum* Hedw., Spec. Musc. 178. 1801. ........ W.H.

– ***palustre*** (Hedw.) Schwaegr., Spec. Musc. Suppl. 3(1): 216. 1827. *Mnium palustre* L. *ex* Hedw., Spec. Musc. 188. 1801.
........ W.H.; Nepal

– ***turgidum*** (Wahlenb.) Schwaegr., Spec. Musc. Suppl. 3(1) [Mon]: 7. 1827. *Mnium turgidum* Wahlenb., Fl. Lapp. 351. 1812.
........ W.H.; Nepal

***Aulacopilum*** Wils., London J. Bot. 7: 90. 1848. (Erpodiaceae)

Type: *A. glaucum* Wils.

– ***abbreviatum*** Mitt., J. Linn. Soc. Bot. 13: 308. 5D. 1873.
........ W.H.; S.I.

– ***luzonense*** Bartr., Philipp. J. Sc. 68: 169. 209. 1939. ........ W.H.

– ***tumidulum*** Thwait. *et* Mitt., J. Linn. Soc. Bot. 13: 307. 5C. 1873.
........ S.I.

***Barbella*** Fleisch. *in* Broth., Nat. Pfl. 1(3): 823. 1906. emend. Nog., J. Hattori Bot. Lab 3: 75. 1948. (Meteoriaceae)

– ***angustifolia*** Gang., Mosses E. India 5: 1338. 1976.
........ E.H. Endemic

var. ***subtenax*** Gang., Mosses E. India 5: 1339. 1976.
........ E.H. Endemic

– ***bombycina*** (Ren. *et* Card.) Fleisch. in Broth., Nat. Pfl. 1(3): 824. 1906. *Meteorium bombycinum* Ren. *et* Card., Bull. Soc. R. Bot. Belg. 38(1): 26. 1900. ........ W.H.; E.H.; Nepal

– ***compressiramea*** (Ren. et Card.) Fleisch. *in* Broth., Nat. Pfl. 1(3): 824. 1906. *Meteorium compressirameum* Ren. et Card., Bull. Soc. R. Bot. Belg. 38(1): 27. 1900. ........ W.H.; E.H.; Nepal

– ***convolvens*** (Mitt.) Broth., Nat. Pfl. 1(3): 824. 1906. *Meteorium convolvens* Mitt., J. Linn. Soc. Bot. Suppl. 1: 90. 1859.
........ E.H.; S.I.

– ***cubensis*** (Mitt.) Broth., Nat. Pfl. 1(3): 824. 1906. *Meteorium cubense* Mitt., J. Linn. Soc.Bot. 12: 435. 1869.
........ E.H.; S.I.; Nepal

– ***determesii*** (Ren. *et* Card.) Fleisch. in Broth., Nat. Pfl. 1(3): 824. 1906. *Meteorium determesii* Ren. *et* Card., Bull. Soc. R. Bot. Belg. 38(1): 25. 1900. ........ Himalayas; S.I.

– ***enervis*** (Thwait. *et* Mitt.) Fleisch. in Broth., Nat. Pfl. 1(3): 824. 1906. *Meteorium enerve* Thwait. *et* Mitt., J. Linn. Soc. Bot. 13: 317. 1873. ........ E.H.; S.I.

– ***pendula*** (Sull.) Fleisch., Musci Fl. Buitenzorg 3: 812. 1908. *Meteorium pendulum* Sull. *in* Gray, Man. Bot. N.U. States ed. 2: 681. 1856. ........ E.H.; S.I.; Nepal

– ***questei*** Card. *et* Dix., Rec. Bot. Surv. India 6(3): 82. 1 f.3. 1914. ........ S.I.

– ***rufifolia*** (Thwait. *et* Mitt.) Broth., Nat. Pfl. 1(3): 824. 1906. *Meteorium rufifolium* Thwait. *et* Mitt., J. Linn. Soc. Bot. 13. 316. 1873. ........ E.H.; S.I.

– ***spiculata*** (Mitt.) Broth., Nat. Pfl. 1(3): 824. 1906. *Meteorium spiculatum* Mitt., J. Linn. Soc. Bot. Suppl. 1: 90. 1859. ........ E.H.; S.I.; Nepal

– ***stevensii*** (Ren. *et* Card.) Fleisch. in Broth., Nat. Pfl. 1(3): 824. 613. 1906. *Meteorium stevensii* Ren *et* Card., Bull. Soc. R. Bot. Belg. 34(2): 72. 1896. ........ E.H.; Nepal

– ***tenax*** (C.Muėll.) Broth., Nat. Pfl. 1(3): 824. 1906. *Neckera tenax* C. Muell., Linnaea 36: 23. 1869. ........ S.I.

– ***turgida*** Nog. in Nog. *et al.*, Musci Fl. East. Himalayas 571. 1966. ........ Nepal, Endemic

***Barbula*** Hedw., Spec. Musc. 115. 1801. (Pottiaceae)

L.Type : *B.unguiculata* Hedw.

– ***arcuata*** Griff., Calcutta J. Nat. Hist. 2: 491. 1842. ........ W.H.; E.H.; S.I.; C.I.; G.P.; Nepal

– ***confertifolia*** Mitt., J. Linn. Soc. Bot. Suppl. 1: 36. 1859. ........ W.H.; Nepal

– ***dharvarensis*** Dix., J. Indian Bot. 2: 182.7. 1921. ........ S.I.

– ***funalis*** Dix. *et* Badhw., Rec. Bot. Surv. India 12(2): 170. 1938. ........ W.H.

– ***fuscescens*** Wallich, Num. List Herb. E. Ind. Comp. n. 7576. 1832. ........ W.H.; Nepal

– ***gracilenta*** Mitt., J. Linn. Soc. Bot. Suppl. 1: 35. 1859. ........ W.H.; E.H.; G.P. Endemic

– ***gregaria*** (Mitt.) Jaeg., Ber. S. Gall. Naturw. Ges. 1871-72: 424. 1873. *Tortula gregaria* Mitt., Musc. Ind. Or. 29. 1859. ........ W.H.; E.H.; Nepal

– ***horricomis*** C. Muell. *ex* Gang., Nov. Hedwigia 12: 423. 1966. ........ W.H.; E.H.

– ***indica*** (Hook.) Spreng. *in* Steud. Nomencl. Bot. 2: 72. 1824. ........ W.H.; E.H.; S.I.; G.P.; R.; Nepal

– ***leucodontoides*** C. Muell. *in* Par. *ex* Gang., Nov. Hedwigia 12: 425. 1966. ........ W.H.; E.H.

– ***marginatula*** C. Muell. *ex* Gang., Nov. Hedwigia 12: 424. 1966. ........ E.H.

– ***microstoma*** (Dix. *et* Badhw.) R.S. Chopra, Bryologist 80: 544. 1977. *Didymodon microstomus* Dix. & Badhw., Rec. Bot. Surv. India 12(2): 169. 1938. ........ W.H.

– ***obscura*** Mitt., J. Linn. Soc. Bot. Suppl. 1:33. 1859. ........ W.H.

– ***ovata*** Mitt., J. Linn. Soc. Bot. Suppl. 1: 36. 1859. ........ W.H.

– ***reflexa*** (Brid.) Brid., Mant. Musc. 93. 1819. *Tortella reflexa* Brid., Spec. Musc.1: 255. 1806. ........ W.H.; E.H.

– ***spathulifolia*** (Dix *et* P. Vard). Zand., Bull. Buffalo Soc. Nat. Sc. NY 32: 1923. *Merceyopsis spathulifolia* Dix. *et* P. Vard., Arch. Bot. 1(8-9): 164. 1927. ........ I.O.

– ***subcontorta*** Broth., Sitzungsber. Ak. Wiss. Wien 133: 56. 1924. ........ W.H.

– ***tenuirostris*** Brid., Bryol. Univ. 1: 826. 1827. ........ E.H.; G.P.; Nepal

– ***tortelloides*** C. Muell., Linnaea 42: 337. 1879. ........ E.H.

– ***unguiculata*** Hedw., Spec. Musc. 118. 1801. ........ W.H.; E.H.

– ***vardei*** R.S. Chopra, Bryologist 80: 544. 1977. ........ I.O.

– ***vinealis*** Brid., Bryol. Univ. 1: 830. 1827. ........ W.H.; E.H.; C.I.; Nepal

***Bartramia*** Hedw., Spec. Musc. 164. 1801. (Bartramiaceae)

L. Type : *B. halleriana* Hedw.

– ***gathica*** Card. *et* P. Vard., Rev. Bryol. 50(2): 22.9.1923. ........ S.I.

– ***halleriana*** Hedw., Spec. Musc. 164. 1801. ........ W.H.; E.H.; Nepal

– ***ithyphylla*** Brid., Musc. Rec. 2(3): 132. 1f. 6. 1803. ........ W.H.; E.H.; Nepal

– ***leptodonta*** Wils., Kew J. Bot. 9: 369. 1857. ........ W.H.; E.H.; S.I. Endemic

– ***madurensis*** Dix. *et* P. Vard., Rev. Bryol. 52: 39. 1925. ........ S.I.

– ***pomiformis*** Hedw., Spec. Musc. 164. 1801. ........ W.H.; E.H.

var. ***elongata*** Turn., Ann. Bot. 1: 527. 1805. ........ W.H.; E.H.

– ***subpellucida*** Mitt., J. Linn. Soc. Bot. Suppl. 1: 59. 1859. ........ W.H.; E.H.; Nepal

– ***subulata*** B.S.G., Bryol. Eur. 4: 53. 315. 1846. ........ W.H.; E.H.; S.I.

***Bartramidula*** Bruch & Schimp. in B.S.G., Bryol. Eur. 4: 55. 1846. (Bartramiaceae)

L. Type : *B. wilsonii* Bruch & Schimp. in B.S.G.

– ***bartramioides*** (Griff.) Wijk *et* Marg., Taxon 7: 289. 1958. *Weisia bartramioides* Griff., Calcutta J. Nat. Hist. 2: 489. 1842. ........ W.H.; E.H.; S.I.

var. ***sikkimensis*** (Kab.) Gang., Mosses E. India 4: 1106. 1974. ........ W.H.; E.H.; S.I.

– ***dispersa*** Card. et P. Vard., Rev. Bryol. 59(2): 23.11. 1923. ........ S.I.

– ***roylei*** (Hook.*f.*) B.S.G., Bryol. Eur. 4: 55. 1846. *Glyphocarpa roylei* Hook.*f.* in Hook., Icon. PI. Rar. 2: 194C. 1837. ........ W.H.; S.I. C.I.; G.P.; Nepal

***Bellibarbula*** Chen, Hedwigia 80: 222. 1941. (Pottiaceae)

Type: *B. kurziana* Chen

– ***kurziana*** Chen, Hedwigia 80: 223. 38 f. 1-5. 1941. ........ E.H.; Nepal

var. ***purpurascens*** Gang., Nov. Hedwigia 8: 148. 1964. ........ E.H.

***Blindia*** Bruch & Schimp. in B.S.G, Bryol. Eur. 2: 17. 1846. (Seligeriaceae)

Type: *B. acuta* (Hedw.) Bruch & Schimp. in B.S.G.

- ***campylopodioides*** Dix. *et* Badhw., Rec. Bot. Surv. India 12(2): 168. 1938. ........ W.H.
- ***himalayana*** Dix. *et* Badhw., Rec. Bot. Surv. India 12(2): 167. 1938. ........ W.H.
- ***roerichii*** Williams, Bryologist 34(4): 49. 3. 1931. ........ W.H.
- ***sordida*** (Mitt.) C.Muell., Bot. Zeit. 22: 349. 1864. *Dicranum sordidum* Wils. *ex* Mitt., J. Linn. Soc. Bot. Suppl. 1: 18. 1859. ........ W.H.; E.H.; Nepal

***Brachymenium*** Schwaegr., Spec. Musc. Suppl. 2(1): 131. 1824. (Bryaceae)

L. Type : *B. nepalense* Hook. in Schwaegr.

- ***acuminatum*** Harv. in Hook., Icon. Pl. Rar. 1: 19. 1836. ........ E.H.; S.I.; Nepal
- ***alpinum*** Ochi, Hikobia 4: 9. 1964. ........ E.H. Endemic
- ***bryoides*** Hook. *ex* Schwaegr., Spec. Musc. Suppl. 2(1): 134. 135. 1824. ........ W.H.; E.H.; S.I.; Nepal
- ***exile*** (Doz. *et* Molk.) Bosch *et* Lac., Bryol. Jav. 1: 139. 1860. *Bryum exile* Doz. et Molk., Ann. Sc. Nat. Bot. ser. 3, 2: 300. 1844. ........ W.H.; E.H.; S.I.; Nepal
- ***indicum*** (Doz. *et* Molk.) Bosch *et* Lac., Bryol. Jav. 1: 141. 1860. *Bryum indicum* Doz. *et* Molk., Musci Fr. Ined. Archip. Indici 1: 22. 1845. ........ G.P.
- ***lanceolatum*** Hook. *f.* et Wils., Fl. Tasman. 2: 188. 173 f.8. 1859. ........ C.I.
- ***longicolle*** Ther., Bull. Soc. Bot. Geneve 26: 85. ic. 1936. ........ E.H.; Nepal
- ***longidens*** Ren. *et* Card., Bull. Soc. R. Bot. Belg. 41(1): 63. 1905. ........ E.H. Endemic
- ***microstomum*** Harv. in Hook., Icon. Pl. Rar. 1: 19 f.4. 1836. ........ E.H.; C.I.; Nepal
- ***nepalense*** Hook. in Schwaegr., Spec. Musc. Suppl. 2(1): 131. 135. 1824. ........ W.H.; E.H.; S.I.; Nepal
- ***ochianum*** Gang., Mosses E. India 4: 934. 1974. ........ E.H.; Nepal

– ***ptychothecium*** (Besch.) Ochi, Adv. Front. Pl. Sc. 4: 108. 1963. *Bryum ptychothecium* Besch., Ann. Sc. Nat. Bot. ser. 7, 15: 66. 1892. ........ E.H.

– ***sikkimense*** Ren. *et* Card., Bull. Soc. R. Bot. Belg. 38(1): 12. 1900. ........ E.H. Endemic

– ***walkeri*** Broth., Rec. Bot. Surv. India 1: 320. 1899. ........ E.H.; S.I. Endemic

***Brachythecium*** Schimp. in B.S.G., Bryol. Eur. 6: 5. 1853. (Brachytheciaceae)

L. Type: *B. rivulare* Schimp.

– ***brachycladum*** (Broth.) Par., Ind. Bryol. Suppl. 42. 1900. *Hypnum brachycladum* Broth., Act. Soc. Sc. Fenn. 24(2): 34. 1898. ........ W.H. Endemic

– ***buchananii*** (Hook.) Jaeg., Ber. S. Gall. Naturw. Ges. 1876-77: 341. 1878. *Hypnum buchananii* Hook., Trans. Linn. Soc. London 9: 320. 28 f. 3. 1808. ........ W.H.; E.H.; S.I.; Nepal

var. ***cuspidiferum*** (Mitt.) Gang. & Vohra in Gang., Mosses E. India 7: 1712. 1978. *Hypnum cuspidiferum* Mitt., J. Linn. Soc. Bot. Suppl. 1: 78. 1859. ........ W.H.; E.H.; S.I.; Nepal

– ***cameratum*** (Mitt.) Jaeg., Ber. S. Gall. Naturw. Ges. 1876-77: 342. 1878. *Hypnum cameratum* Mitt., Musci Ind. Or.: 78. 1859. ........ W.H.; E.H.

– ***campestre*** (C.Muell.) B.S.G., Bryol. Eur. 6: 16. 545. 1853. *Hypnum rutabulum* Hedw., var. *campestre* C.Muell., Syn. 2: 368. 1851. ........ W.H.

– ***chakratense*** Vohra, Bull. Bot. Surv. India 22: 118. 1980. ........W.H.

– ***cirrhosum*** (Schwaegr.) Schimp., Syn. 696. 1860. *Hypnum cirrhosum* Schwaegr. in Schultes, Reise Glockner 365. 1804. ........ W. H.; Nepal

– ***curvatulum*** (Broth.)Par., Ind. Bryol. Suppl. 43. 1900. *Hypnum curvatulum* Broth., Act. Soc. Sc. Fenn. 24(2): 38. 1899. ........ W.H. Endemic

– ***falcatulum*** (Broth.) Par., Ind. Bryol. Suppl. 43. 1900. *Hypnum falcatulum* Broth., Act. Soc. Sc. Fenn. 24(7): 34. 1899. ........ W.H.

– ***formosanum*** Tak., Hattori Bot. Lab. 15: 2. 8. 1955.
........ E.H.; Nepal

– ***garhwalense*** Vohra, Bull. Bot. Surv. India 22: 120. 1980.
........ W.H. Endemic

– ***glaciale*** B.S.G., Bryol. Eur. 6: 15. 542. 1853. ........ W. H.

– ***indico-populeum*** Dix., Rec. Bot. Surv. India 9(4): 312. 1 f. 6. 1926.
........ W.H. Endemic

– ***kamounense*** (Harv.) Jaeg., Ber. S. Gall. Naturw. Ges. 1876-77: 342. 1878. *Hypnum kamounense* Harv. in Hook., Ic. Pl. Rar. 1: 24 f.10. 1836. ........ W.H.; E.H.; Nepal

– ***kashmirense*** (Broth.) Par., Ind. Bryol. Suppl. 45. 1900. *Hypnum kashmirense* Broth., Act. Soc. Sc. Fenn. 24(2): 37. 1898.
........ W.H. Endemic

– ***laevi-velutinum*** Dix., Ann. Bryol. 3: 68. 1929. ........ W.H. Endemic

– ***laxifolium*** Dix., Not. R. Bot. Gard. Edinburgh 19: 296. 1938.
........ W.H. Endemic

– ***longicuspidatum*** (Mitt.) Jaeg., Ber. S. Gall. Naturw. Ges. 1876-77: 341. 1878. *Hypnum longicuspidatum* Mitt., J. Linn. Soc. Bot. Suppl. 1: 77. 1859. ........ W.H.; E.H. Endemic

– ***microsericeum*** Dix. in Blatt. *et* Fern., J. Indian Bot. Soc. 10: 153. 1931. ........ W.H.; E.H. Endemic

– ***myurelliforme*** Dix., Ann. Bryol. 3: 69. 1930. ........ W.H. Endemic

– ***obsoletinerve*** Dix., Rec. Bot. Surv. India 9(4): 312. If. 5. 1926.
........ W.H. Endemic

– ***pachythecum*** (Dix.) Vohra, Bull. Bot. Surv. India 21: 171. 1979. *Campylium pachythecum* Dix., Not. Roy. Bot. Gard. Edinburgh 19: 295. 1938. ........ I.O. Endemic

– ***plumosum*** (Hedw.) B.S.G., Bryol. Eur. 6: 8. 537. 1853. *Hypnum plumosum* Hedw., Spec. Musc. 257. 1801.
........ W.H.; E.H.; S.I.; Nepal

– ***populeum*** (Hedw.) B.S.G., Bryol. Eur. 6: 7. 535. 1853. *Hypnum populeum* Hedw., Spec. Musc. 270. 70 f. 1-6. 1801.
........ W.H.; Nepal

– ***procumbens*** (Mitt.) Jaeg., Ber. S. Gall. Naturw. Ges. 1876-77: 341. 1878. *Hypnum procumbens* Mitt., J. Linn. Soc. Bot. Suppl. 1: 79. 1859. ........ W.H.; E.H.; S.I.; Nepal

– ***reflexum*** (Stark.) B.S.G., Bryol. Eur. 6: 12. 539. 1853. *Hypnum reflexum* Stark. in Web. *et* Mohr, Bot. Taschenb. 306. 476. 1807. ........ I.O.

– ***rivulare*** B.S.G., Bryol. Eur. 6: 17. 546. 1853. ........ W.H.; E.H.; Nepal

var. ***alare*** (Dix.) Vohra, Bull. Bot. Surv. India 21: 171. 1979. *Cirriphyllum alare* Dix., Not. R. Bot. Gard. Edinburgh 19: 297. 1938. ........ W.H. Endemic

– ***rutabulum*** (Hedw.) B.S.G., Bryol. Eur. 6: 15. 543. 1853. *Hypnum rutabulum* Hedw., Spec. Musc. 276. 1801. ........ W.H.; Nepal

– ***salebrosum*** (Web. *et* Mohr) B.S.G., Bryol. Eur. 6: 20. 549. 1853. *Hypnum salebrosum* Hoffm. *ex* Web. *et* Mohr, Bot. Taschenb. 312. 1807. ........ W.H.; Nepal

– ***spurio-populeum*** (Broth.) Par., Ind. Bryol. Suppl. 50. 1900. *Hypnum spurio-populeum* Broth., Act. Soc. Sc. Fenn. 24(7): 238. 1899. ........ W.H. Endemic

– ***starkei*** (Brid.) B.S.G.

var. ***explanatum*** (Brid.) Moenk., Laubm. Eur. 819. 1927. *Hypnum rutabulum* Hedw., var. *explanatum* Brid., Spec. Musc. 2; 184. 1812. ........ W.H.

– ***stricticalyx*** Dix. *et* Badhw., Rec. Bot. Surv. India 12 : 178. 1938. ........ W.H. Endemic

– ***velutinum*** (Hedw.) B.S.G., Bryol. Eur. 6: 9. 538. 1853. *Hypnum velutinum* Hedw., Spec. Musc. 272. 1801. ........ W.H.

– ***waziriense*** Dix., J. Bombay Nat. Hist. Soc. 33(2): 283. 1929. ........ W.H.

– ***wichurae*** (Broth.) Par., Ind. Bryol. Suppl. 52. 1900. *Hypnum wichurae* Broth., Hedwigia 38: 239. 1900. ........ E.H.; Nepal

***Braunia*** B.S.G., Bryol. Eur. 3: 159. 1846. (Hedwigiaceae)

Type: *B. sciuroides* (Bals. *et* De Not) B.S.G.

– ***apiculata*** Card. *in* Foreau, J. Bombay Nat. Hist. Soc. 58(1): 29. 1961. ........ S.I.

– ***attenuata*** (Mitt.) Jaeg., Ber. S. Gall. Naturw. Ges. 1869-70: 172. 1871. *Hedwigia attenuata* Mitt., J. Linn. Soc. Bot. Suppl. 1: 123. 1859. ........ W.H.

– ***secunda*** (Hook.) B.S.G., Bryol. Eur. 3: 161. 1846. *Hedwigia secunda* Hook., Musci Exot. 1: 46. 1818. ........ S.I.

***Breutelia*** (Bruch & Schimp.in B.S.G.) Schimp., Coroll. 85. 1856. (Bartramiaceae)

Type: *B. arcuata* (Sw.) Schimp.

– ***arundinifolia*** (Dub.) Fleisch., Musci Fl. Buitenzorg 2: 630. 120. 1904. *Bartramia arundinifolia* (Dub.) C. Muell., Gen. Musc. Fr. 347. 1900. ........ I.O.

– ***dicranacea*** (C.Muell.) Mitt., J. Linn. Soc. Bot. Suppl. 1: 64. 1859. *Bartramia dicranacea* C. Muell., Bot. Zeit. 11: 57. 1853. ........ W.H.; E.H.; S.I.; Nepal

– ***sclerodictya*** Card. in Grand., Hist. Madag. 39: 315. 1915. ........ S.I.

***Brothera*** C. Muell., Gen. Musc. Fr. 258. 1900. (Dicranaceae)

Type: *B. ankerkronae* C. Muell.

– ***capillifolia*** Dix., J. Bombay Nat. Hist. Soc. 39(4): 774. 1 f.1. 1937. ........ E.H.

– ***himalayana*** Broth., Nat. Pfl. 1(3): 330. 1901. ........ W.H.; E.H. Endemic

– ***leana*** (Sull.) C. Muell., Gen. Musc. Fr. 259. 1900. *Leucophanes leanum* Sull., Musc. Allegh. 41. 1846. ........ W.H.; E.H.; S.I.; Nepal

***Brotherella*** Loesk. *ex* Fleisch., Nov. Guinea 12(2): 119. 1914. (Sematophyllaceae)

Type: *B. lorentziana* (Mol.) Loesk.

– ***amblystegia*** (Mitt.) Broth., Nat. Pfl. ed. 2, 11: 425. 1925. *Stereodon amblystegia* Mitt., Musc. Ind. Or. 97. 1859. ........ E.H. Endemic

– ***curvirostris*** (Schwaegr.) Fleisch., Nov. Guinea 12(2): 120. 1914.*Neckera curvirostris* Schwaegr., Spec. Musc. Suppl. 3(1): 230 b 1828. ........ E.H.; Nepal

– ***dixonii*** Herz., Ann. Bryol. 12: 95. 14. 1939. ........ E.H. Endemic

– ***erythrocaulis*** (Mitt.) Fleisch., Musci Fl. Buitenzorg 4: 1245. 1923. *Stereodon erythrocaulis* Mitt., Musci Ind. Or. 97. 1859. ........ E.H.

– ***falcata*** (Doz. *et* Molk.) Fleisch., Nov. Guinea 12(2): 120. 1914. *Leskea falcata* Doz. *et* Molk., Ann. Sc. Nat. Bot. ser. 3, 2: 310. 1844. ........ E.H.

– ***filiformis*** Dix., J. Bombay Nat. Hist. Soc. 39(4): 792. 1937. ........ E.H. Endemic

– ***harveyana*** (Mitt.) Dix., Not. R. Bot. Gard. Edinburgh 19: 298. 1938. *Stereodon harveyanus* Mitt., Musci Ind. Or. 102. 1859. ........ W.H.; Nepal

– ***nictans*** (Mitt.) Broth., Nat. Pfl; ed. 2, 11: 425. 1925. *Stereodon nictans* Mitt., Musci. Ind. Or. 98. 1859. ........ E.H.; Nepal

– ***pallida*** (Ren. *et* Card.) Fleisch., Nov. Guinea 12(2): 120. 1914. *Acanthocladium pallidum* Ren. *et* Card., Bull. Soc. R. Bot. Belg. 41(1)93. 1905. ........ E.H. Endemic

– ***perpinnata*** (Broth.) Fleisch., Nov. Guinea 12(2): 120. 1914. *Stereodon perpinnatus* Broth., Nat. Pfl. 1(3): 1068.764. 1908. ........ W.H.; E.H.; Nepal

– ***propinqua*** (Harv.) Fleisch., Musci Fl. Buitenzorg 4: 1245. 1923. *Hypnum propinquum* Harv. in Hook., Icon. P1. Rar. 1: 24. 1836. ........ E.H.; Nepal

– ***yokahamae*** (Broth.) Broth., Nat. Pfl. ed. 2, 11: 425. 1925. *Stereodon yokohamae* Broth., Hedwigia 38: 235. 1899. ........ W.H.

***Bryhnia*** Kaur., Bot. Not. 1892: 61. 1892. (Brachytheciaceae)

Type: *B. scabrida* (Lindb.) Kaur.

– ***decurvans*** (Mitt.) Dix., J. Bot. 60: 287. 1922. *Hypnum decurvans* Mitt., J. Linn. Soc. Bot. Suppl. 1: 78. 1859. ........ W.H.; E.H.; Nepal

– ***nepalensis*** Tak., Bull. Nat. Sc. Mus. Tok. 9: 374. 1966. ........ Nepal

– ***novae-angliae*** (Sull. *et* Lesq.) Grout, Bull. Torr. Bot. Cl. 25: 229. 1898. *Hypnum novae-angliae* Sull. *et* Lesq., Musci Bor. Am. 1: 73. 1856. ........ E.H.; Nepal

***Bryoerythrophyllum*** Chen, Hedwigia 80: 4. 1941. (Pottiaceae)

L. Type: *B. recurvirostrum* (Hedw.) Chen

– ***alpigenum*** (Vent.) Chen, Hedwigia 80: 5. 257.53. 1941. *Didymodon alpigenus* Vent. in Jur., Laubmfl. Oest. Ungarn 98. 1882. ........ W.H.

– ***atrorubens*** (Besch.) Chen, Hedwigia 80: 262. 1941. *Trichostomum atrorubens* Besch., Rev. Bryol. 18: 88. 1891. ........ W.H.; Nepal

– ***ferrugineum*** Gang., Nov. Hedwigia 8: 147. 1964. ........ E.H. Endemic

– ***gymnostomum*** (Broth.) Chen, Hedwigia 80: 5. 255.51 f.1-4. 1941. *Didymodon gymnostomus* Broth., Symb. Sin. 4: 39. 1929. ........ W.H.; E.H.

– ***inaequalifolium*** (Tayl.) Zand. Bull. Buffalo Soc. Nat. Sc. NY 32: 112. 1993. *Barbula inaequalifolia* Tayl., London J. Bot. 5: 49. 1846.

– ***recurvirostrum*** (Hedw.) Chen, Hedwigia 80: 5. 255. 52 f.1-2. 1941. *Wiesia recurvirostris* Hedw., Spec. Musc. 71. 1801. ........ W.H.; S.I.; Nepal

– ***wallichii*** (Mitt.) Chen, Hedwigia 80:5.260.54. 1941. *Desmatodon wallichii* Mitt., Musci Ind. Or. 38. 1859. ........ W.H.; E.H.; Nepal

– ***yunnanense*** (Herz.) Chen, Hedwigia 80: 5.259.52 f. 3-5. 1941. *Erythrophyllum yunnanense* Herz., Hedwigia 65: 152. 1925. ........ W.H.; E.H.

var. ***noguchianum*** Gang., Mosses E. India 3: 750. 1972. ........ W.H.; E.H.

***Bryosedgwickia*** Card. *et* Dix., J. Bot. 50: 154. 1912. (Hypnaceae)

L. Type: *B. kirtikarii* Card. *et* Dix.

– ***aurea*** (Schwaegr.) Fleisch., Hedwigia 63: 211. 1922. *Neckera aurea* Schwaegr., Spec. Musc. Suppl. 3(1): 217 b. 1827. ........ W.H.; E.H.; Nepal

– ***kirtikarii*** Card. *et* Dix., J. Bot. 50: 154. 517 f.3. 1912. ...S.I. Endemic

– ***polyantha*** (Hedw.) B.S.G., Bryol. Eur. 5: 88. 445. 1851. ........ W.H.

***Bryum*** Hedw., Spec. Musc. 178. 1801 emend Schimp., Syn. ed. 1. 1860 (Bryaceae)

L. Type: *B. argenteum* Hedw.

– ***alpinum*** Huds. *ex* With., Syst. Arr. Brit. Pl. ed. 4, 3: 824. 1801.
........ W.H.; E.H.; S.I.; Nepal

var. ***mildeanum*** (Jur.) Podp. Consp. 357. 1954. *B. mildeanum* Jur., Verh. Zool. Bot. Ges. Wien 12: 967. 1862. ......... W.H.

– ***apiculatum*** Schwaegr., Spec. Musc. Suppl. 1(2): 102. pl. 72. 1816.
........ W.H.; E.H.; S.I.; G.P.; Nepal

– ***argenteum*** Hedw., Spec. Musc. 181. 1801.
........ W.H.; E.H.; R.; C.I.; S.I.; Nepal

var. ***lanatum*** (P. Beauv.) Hamp., Linnaea 13: 44. 1839. *et* B.S.G. Bryol. Eur. 4: 148. 1839. *Mnium lanatum* P. Beauv. Prodr. 75. 1805.
........ W.H.; E.H.; C.I.; S.I.; Nepal

– ***atrovirens*** Vill. *ex* Brid, Musci Rec. 2(3): 48. 1803.
........ W.H.; E.H.

– ***badhwari*** Ochi, Flora E. Himalaya, Univ. Tokyo : 557. 1966.
........ W.H.; E.H. Endemic

– ***bessonii*** Ren. *et* Card., Prodr. F1. Bryol. Madag. Suppl. 59. 1909.
........ S.I.; Nepal

– ***bicolor*** Dicks., P1. Crypt. Brit. fasc. 4: 16. 1801.
........ W.H.; E.H.; R.; Nepal

– ***billardieri*** Schwaegr., Spec. Musc. Suppl. 1(2): 115. 76. 1816.
........ W.H.; E.H.; S.I.; Nepal

– ***bornholmense*** Wink. & R. Ruth., Hedwigia 38 (Beibl.3): 120. 1899.
........ R.

– ***bryoides*** (R. Brown) Wijk *et* Marg., Taxon 8: 72. 1959. *Pohlia bryoides* R. Brown, Suppl. App. Capt. Parry's Voyage 296. 1823.
........ W.H.

– ***caespiticium*** L. *ex* Hedw., Spec. Musc. 1801. 1801.
........ W.H.; E.H.; Nepal

– ***capillare*** L.*ex* Hedw., Spec. Musc. 182. 1801.
........ W.H.; E.H.; S.I.; Nepal

– ***cellulare*** Hook. *in* Schwaegr., Spec. Musc. Suppl. 3(1): 214a. 1827.
........ W.H.; E.H.; Nicobar Is.; Nepal

– ***coronatum*** Schwaegr., Spec. Musc. Suppl. 1(2): 103. 71. 1816.
........ W.H.; E.H.; S.I.; R.; C.I.; G.P.; Andaman Is.; Nepal

– ***dunense*** A.J.E. Sm. & H. Whitehouse R. ........ R.

– ***flaccum*** Wils. *ex* Mitt., J. Linn. Soc. Bot. Suppl. 1: 72. 1859. ........ E.H. Endemic

– ***heterophyllum*** Card. in Dix. *ex* Gang., Mosses E. India 4: 976. 1974. ....... S.I.; Nepal

– ***klinggraeffii*** Schimp. *in* Klinggr., Höh. Crypt. Preuss. 81. 1858. ........ R.; G.P.

– ***medianum*** Mitt., J. Linn. Soc. Bot. Suppl. 1: 74. 1859. ........ W.H.; E.H.; S.I. Endemic

– ***muehlenbeckii*** B.S.G., Bryol. Eur. 4: 163. 381. 1846. ........ W.H.; E.H.; S.I. Endemic

ssp. ***evanidinerve*** (Broth.) Podp., Act. Ac. Sc. Cech. Bas. Brun. 27: 120. 1955. *Bryum evanidinerve* Broth., Act. Soc. Sc. F. Fl. Fenn. 24(2): 24. 1899. ........ W.H.

– ***paradoxum*** Schwaegr., Spec. Musc. Suppl. 3(1): 224a. 1827. ........ W.H.; E.H.; S.I.; Nepal

var. ***reflexifolium*** (Ochi) Ochi, Hikobia 5: 158. 1969. *Bryum pseudoalpinum* var. *reflexifolium* Ochi, J. Jap. Bot. 39: 52.1964. ........ E.H.; Nepal

– ***plumosum*** Doz. *et* Molk., Ann. Sc. Nat. Bot. ser. 3(2): 301. 1844. ........ W.H.; E.H.; G.P.; S.I.; Nepal

– ***porphyroneuron*** C.Muell., Bot. Zeit. 11: 22. 1853.

var. ***erythrinum*** (Mitt.) Fleisch. *in* Broth., Nat. Pfl. 1(3): 590. 1904. *Bryum erythrinum* Mitt., Musci Ind. Or. 70. 1895. .......W.H.; E.H.; S.I.

– ***pseudotriquetrum*** (Hedw.) Schwaegr., Spec. Musc. Suppl. 1(2): 110. 1816. *Mnium pseudotriquetrum* Hedw., Spec. Mus. 190. 1801. ....... W.H.; E.H.; Nepal

– ***recurvulum*** Mitt., J. Linn. Soc. Bot. Suppl. 1: 74. 1859. ....... W.H.; Nepal

– ***retusifolium*** Card. *et* Vard., Rev. Bryol. 50: 20. 7. 1923 ....... S.I.; Nepal

– ***rubens*** Mitt., Kew J. Bot. 8: 232. 1856. ....... R.

– ***salakense*** Card., Annuair. Cons. Jard. Bot. Geneve 15-16: 166. 1912. ....... I.O.

– ***tuberosum*** Mohamed & Damanhri, Brylogist 93(3): 288.1990. ....... S.I

– ***turbinatum*** (Hedw.) Turn., Musc. Hib. 127. 1804. *Mnium turbinatum* Hedw., Spec. Musc. 191. 1801. ....... W.H.

– ***uliginosum*** (Brid.) B.S.G., Bryol. Eur. 4: 88. 339. 1839. *Cladodium uliginosum* Brid., Bryol. Univ. 1: 841. 1827. ....... W.H.; E.H.

– ***weigelii*** Spreng., Mant. Prim. Fl. Halens. Add. 55. 1807. ....... W.H.

– ***wightii*** Mitt., J. Linn. Soc. Bot. Suppl. 1: 74. 1859. ....... S.I.; Nepal

***Buxbaumia*** Hedw., Spec. Musc. 166. 1801. (Buxbaumiaceae)

Type: *B. aphylla* Hedw.

– ***himalayensis*** Udar, Sriv. *et* Kumar, Trans. Brit. Bryol. Soc. 6: 266. 1971. ....... W.H. Endemic

***Callicostella*** (C.Muell.) Mitt., J. Linn. Soc. Bot. Suppl. 1: 136. 1859. (Hookeriaceae)

Type: *C. papillata* (Mont.) Mitt.

– ***papillata*** (Mont.) Mitt., J. Linn. Soc. Bot. Suppl. 1: 136. 1859. *Hookeria papillata* Mont., Lond. J. Bot. 3: 632. 1844. ....... E.H.; S.I.; Andaman Is.

***Calliergon*** (Sull.) Kindb., Canad. Rec. Sc. 6(2): 72. 1894. (Amblystegiaceae)

L. Type : *C.cordifolium* (Hedw.) Kindb.

– ***cordifolium*** (Hedw.) Kindb., Canad. Rec. Sc. 6(2): 72. 1894. *Hypnum cordifolium* Hedw., Spec. Musc. 254. 1801. ........ Nepal

– ***nubigenum*** (Mitt.) Broth., Nat. Pfl. 1(3): 1037. 1907. *Stereodon nubigena* Mitt., J. Linn. Soc. Bot. Suppl. 1: 109. 1859. ........ E.H. Endemic

– ***sarmentosum*** (Wahlenb.) Kindb., Canad. Rec. Sc. 6(2): 72. 1894. *Hypnum sarmentosum* Wahlenb., Fl. Lapp. 380. 1812. ........ Nepal

***Calliergonella*** Loesk., Hedwigia 50: 248. 1911. (Amblystegiaceae)

Type: *C.cuspidata* (Hedw.) Loesk.

– ***cuspidata*** (Hedw.) Loesk., Hedwigia 50: 248. 1911. *Hypnum cuspidatum* L. *ex* Hedw., Spec. Musc. 254. 1801.
........ E.H.; Nepal

***Calymperes*** Sw. *in* Web., Tab. Exh. Calyptr. Operc. Gen.3. 1813 (1814). (Calymperaceae)

L. Type : *C. lonchophyllum* Schwaegr.

– ***andamense*** Besch., Ann. Sc. Nat. Bot. ser. 8, 1: 272, 276. 1896.
........ Andaman Is. Endemic

– ***delessertii*** Besch., Ann. Sc. Nat. Bot. ser. 8, 1 : 265, 282. 1896.
........ Andaman Is.

– ***griffithii*** C. Muell. in Besch., Ann. Sc. Nat. Bot. ser. 8, 1: 269, 285. 1896. ........ Nepal

– ***hampei*** Doz. *et* Molk., Bryol. Jav. 1: 48.39. 1856. ........ E.H.

– ***heterophyllum*** (Mitt.) Besch., Ann. Sc. Nat. Bot. ser. 8, 1: 272, 286. 1896. *Syrrhopodon heterophyllum* Mitt., Musc. Ind. Or. 40. 1859. ........ Nepal

– ***kurzianum*** Hamp. *ex* C. Muell., Flora 61: 82. 1878.
........ Andaman Is. Endemic

– ***linguatum*** C.Muell. *ex* Besch., Ann. Sc. Nat. Bot. ser. 8, 1: 265, 290. 1896. ........ Andaman Is. Endemic

– ***manii*** C. Muell. *ex* Besch., Ann. Sc. Nat. Bot. ser. 8, 1: 267, 291. 1896. ........ S.I.; Andaman Is.

– ***nicobarense*** Hamp. in Besch., Ann. Sc. Nat. Bot. ser. 8, 1: 269, 294. 1896. ........ Nicobar Is. Endemic

– ***noakhalensis*** Bruehl *et* Sarkar, J. Dept. Sc. Cal. Univ. 10: 3. 1929.
........ I.O.

– ***punctulatum*** Hamp. in Besch., Ann. Sc. Nat. Bot. ser. 8, 1: 296, 300. 1896. ........ Nicobar Is.

– ***sikkimense*** Broth. *ex* Gang., Nov. Hedwigia 12: 421-422. 1966.
........ E.H. Endemic

– ***sundarbanense*** Gang., Mosses E. India 3: 611. 1972.
........ G.P. Endemic

– ***tenerum*** C.Muell., Linnaea 37: 174. 1872.

var. ***teniolata*** Gang., Mosses E. India 3: 600. 1972.
........ G.P.; S.I.

– ***vriesei*** Besch., Ann. Sc. Nat. Bot. ser. 8, 1: 268, 307. 1896. ........ E.H.

***Calymperopsis*** (C. Muell.) Fleisch., Biblioth. Bot. 80: 5. 1913. (Calymperaceae)

– ***semiliber*** (Mitt.) Fleisch., Biblioth. Bot. 80: 5.1913. *Calymperes semi-liberum* Mitt., J. Linn. Soc. Bot. Suppl. 1: 41. 1859. ........ S.I.

***Calyptothecium*** Mitt., J. Linn. Soc. Bot. 10: 190. 1868. (Pterobryaceae)

Type: *C. praelongum* Mitt.

– ***crispulum*** (Bosch et Lac.) Broth., Nat. Pfl. 1(3): 839. 1906. *Neckera crispula* Bosch *et* Lac., Bryol. Jav. 2: 66. 184. 1863. ........ Nepal

– ***dixonii*** Gang., Mosses E. India 5: 1371. 1976. ........ E.H. Endemic

– ***himantocladioides*** Bartr., Philipp. J. Sc. 68: 238, 296. 1939. ........ E.H.

– ***hookeri*** (Mitt.) Broth., Nat. Pfl. 1(3): 839. 1906. *Meteorium hookeri* Mitt., J. Linn. Soc. Bot. Suppl. 1: 86. 1859. ........ W.H.; E.H.; Nepal

– ***nitidum*** (Ren. *et* Card.) Fleisch., Hedwigia 45: 62. 1905. *Garovaglia nitida* Ren. *et* Card., Bull. Soc. R. Bot. Belg. 41(1): 70. 1905. ........ E.H. Endemic

– ***pinnatum*** Nog., Trans. Nat. Hist. Soc. Formosa 24: 417. 3. 1934. ........ E.H.; Nepal

– ***prainii*** Gang., Mosses E. India 5: 1378. 1976. ........ E.H. Endemic

– ***urvilleanum*** (C.Muell.) Broth., Nat. Pfl. 1(3): 839. 1906. *Neckera urvilleana* C. Muell., Syn. 2: 52. 1850. ........ E.H.; S.I.; Nepal

– ***wightii*** (Mitt.) Fleisch., Hedwigia 45: 62. 1905. *Meteorium wightii* Mitt., J. Linn. Soc. Bot. Suppl. 1: 85. 1859. ........E.H.; Nepal

***Camptothecium*** B.S.G., Bryol. Eur. 6: 31. 1853. (Brachytheciaceae)

L. Type : *C. lutescens* (Hedw.) B.S.G.

– ***lutescens*** (Hedw.) B.S.G., Bryol. Eur. 6: 36. 558. 1853. *Hypnum lutescens* Hedw., Spec. Musc. 274. 1801. ........ W.H.

***Campyliadelphus*** (Kindb.) R.S. Chopra, Taxon. Indian Mosses: 442. 1975. (Amblystegiaceae)

– ***elodes*** (Lindb. *in* Hartm.) R.S. Chopra, Taxon. Indian Mosses: 443. 1975. *Amblystegium elodes* Lindb. in Hartm., Handb. Skand. Fl. ed. 10.2:3. 1871. ........ W.H.

– ***protensum*** (Brid.) R.S. Chopra, Taxon. Indian Mosses: 443. 1975. *Hypnum protensum* Brid., Musc. Rec. 2(2): 85. 1801. ........ W.H.

***Campylium*** (Sull.) Mitt., J. Linn. Soc. Bot. 12: 631. 1869. (Amblystegiaceae)

L. Type : *C. hispidulum* (Brid.) Mitt.

– ***chrysophyllum*** (Brid.) J. Lange, Nomencl. Fl. Dan. 210. 1887. *Hypnum chrysophyllum* Brid., Musc. Rec. 2(2): 84. 1801. ........ W.H.

– ***gollanii*** C. Muell. *ex* Vohra, Bull. Bot. Surv. India 12(1-4): 101. 1970. ........ W.H. Endemic

– ***halleri*** (Hedw.) Lindb., Musci Scand. 38. 1879. *Hypnum halleri* Sw. *ex* Hedw., Spec. Musc. 279. 1801. ........ W.H.; E.H.; Nepal

– ***lacerulum*** (Mitt.) Broth., Nat. Pfl. 1(3): 1042. 1908. *Stereodon lacerulus* Mitt., J. Linn. Soc. Bot. Suppl. 1: 115. 1859. ........ E.H.; Nepal

– ***sommerfeltii*** (Myr.) J. Lange, Nomencl. Fl. Dan. 210. 1887. *Hypnum sommerfeltii* Myr., K. Svenska Vetensk Akad. Aorsb. 328. 1831. ........ W.H.

***Campylodontium*** Schwaegr., Spec. Musc. Suppl. 3(1): 211. 1827. (Entodontaceae)

Type: *C. hypnoides* Schwaegr.

– ***flavescens*** (Hook.) Bosch et Lac., Bryol. Jav. 2: 128. 1865. *Pterogonium flavescens* Hook., Musci Exot. 2: 153. 1819. ........ E.H.; S.I.; G.P.; Nepal

– ***perplicatum*** (Ther. *et* P. Vard.) Broth., Nat. Pfl. ed. 2, 11: 534. 1925. *Entodon perplicatus* Ther. *et* P. Vard., Rev. Bryol. 50: 75. 1923. ........ S.I.

***Campylopodiella*** Card., Bull. Herb. Boiss. ser. 2, 8: 90. 1908. (Dicranaceae)

Type: *C. tenella* Card.

– ***ditrichoides*** Nog., Candollea 19 : 170. 1964. ........ W.H.; Nepal

– ***tenella*** Card., Bull. Herb. Boiss. ser. 2, 8: 90. 1908. ........ E.H.; Nepal

***Campylopodium*** (C.Muell.) Besch., Ann. Sc. Nat. Bot. ser. 5, 18: 189. 1873. (Dicranaceae)

Type: *C.euphorocladum* (C.Muell.) Besch.

– ***griffithii*** (Mitt.) Mitt. *ex* Broth., Nat. Pfl. ed. 2,10: 183. 1924. *Leptotrichum griffithii* Mitt., Musc. Ind. Or. 9. 1859. ........ W.H.; E.H.; S.I.; Nepal

– ***khasianum*** (Griff.) Par., Ind. Bryol. 237. 1894. *Dicranum khasianum* Griff., Calcutta J. Nat. Hist. 2: 496. 1842. ........ E.H.; S.I.; Nepal

– ***phascoides*** (C. Muell.) Par., Ind. Bryol. 237. 1894. *Aongstroemia phascoides* C. Muell., Bot. Zeit. 11: 39. 1853. ........ S.I.

***Campylopus*** Brid., Mant. Musc. 71. 1819. (Dicranaceae)

L.Type: *C. flexuosus* (Hedw.) Brid.

– ***albescens*** (C.Muell.) Jaeg., Ber. S. Gall. Naturw. Ges. 1870-71: 429. 1872. *Dicranum albescens* C.Muell., Bot. Zeit. 11: 36. 1853. ........ S.I.; Nepal

– ***alpigena*** Broth., Sitzungsber. Ak. Wiss. Wien Math. Nat. Kl. 133: 562. 1924. ........ E.H.; Nepal

– ***aureus*** Bosch *et* Lac., Bryol. Jav. 1:80.67. 1858. ........ S.I.; Nicobar Is.

– ***durelli*** Broth. *ex* Gang., Nov. Hedwigia 8: 146. 1964. ........ S.I.; Nicobar Is.

– ***ericoides*** (Griff.) Jaeg., Ber. S. Gall. Naturw. Ges. 1870-71. 424. 1872. *Dicranum ericoides* Griff., Calcutta J. Nat. Hist. 2: 499. 1842. ........ E.H.; Nepal

– ***flexuosus*** (Hedw.) Brid., Mant. Musc. 4: 71. 1819. *Dicranum flexuosum* Hedw., Spec. Musc. 145. 1801. ........ Nepal

– ***fragilis*** (Brid.) B.S.G., Bryol. Eur. 1: 164. 90. 1847. *Dicranum fragile* Brid., J.F. Bot. 1800(2): 296. 1801. ........W.H.; E.H.

var. ***pyriformis*** (Schultz.) Agst. Ned. Kruidk. Arch. 57: 332. 8. 1950. *Dicranum pyriformis* Schultz., Prodr. Fl. Starg. Suppl. 73. 1819. ........ W.H.; E.H.

– ***goughii*** (Mitt.) Jaeg., Ber. S.Gall. Naturw. Ges. 1870-71: 424. 1872. *Dicranum goughii* Mitt., Musc. Ind. Or. 17. 1859. ........ W.H.; E.H.; S.I.; Nepal

– ***gracilis*** (Mitt.) Jaeg., Ber. S. Gall. Naturw. Ges. 1870-71: 427. 1872. *Dicranum gracile* Mitt., Musc. Ind. Or. 17. 1859. ........ W.H.; E.H.; S.I.; Nepal

– ***introflexus*** (Hedw.) Brid., Mant. Musc. 72. 1819. *Dicranum introflexum* Hedw., Spec. Musc.147. 1801. .......E. H.; S.I.

– ***involutus*** (C.Muell.) Jaeg., Ber. S.Gall. Naturw. Ges. 1870-71: 418. 1872. *Dicranum involutum* C. Muell., Bot. Zeit. 11: 34. 1853. ........ E.H.; S.I.

– ***laetus*** (Mitt.) Jaeg., Ber. S. Gall. Naturw. Ges. 1870-71: 417. 1872. *Dicranum laetum* Mitt., Musc. Ind. Or. 19. 1859. ........ E.H.; S.I.; Nepal

– ***latinervis*** (Mitt.) Jaeg., Ber. S. Gall. Naturw. Ges. 1870-71: 426. 1872. *Dicranum latinervis* Mitt., Musc. Ind. Or. 17. 1859. ........ E.H.; S.I.; Nepal

– ***milleri*** Ren. *et* Card., Bull. Soc. R. Bot. Belg. 41(1): 17. 1905. ........ E.H. Endemic

– ***richardii*** Brid., Mant. Musc. 73. 1819. ....... W.H.; E.H.; S.I.; Nepal

– ***schwartzii*** Schimp., Musci Eur. Nov. Bryol. Eur. Suppl. fasc. 1-2. 1: 1. 1864. ........ E.H.; Nepal

– ***subfragilis*** Ren. *et* Card., Bull. Soc. R. Bot. Belg. 34(2): 59. 1896. ........ E.H.; S.I.; Nepal

– ***subgracilis*** Ren. et Card. ex Gang., Bull. Bot. Soc. Bengal 14: 26. 1960. ........ E.H.; S.I.; Nepal

– ***subluteus*** (Mitt.) Jaeg., Ber. S. Gall. Naturw. Ges. 1870-71: 417. 1872. *Dicranum subluteum* Mitt., Musc. Ind. Or. 19. 1859. ........ E.H.

***Ceratodon*** Brid., Bryol. Univ. 1: 480. 1826. (Ditrichaceae)

L.Type : *C. purpureus* (Hedw.) Brid.

– ***purpureus*** (Hedw.) Brid., Bryol. Univ. 1: 480. 1826. *Dicranum purpureum* Hedw., Spec. Musc. 136. 1801.
........ W.H.; E.H.; S.I.; Nepal

– ***stenocarpus*** Bruch *et* Schimp. *ex* C. Muell., Syn. 1: 647. 1849.
........ W.H.; E.H.; S.I.; Nepal

***Chaetomitriopsis*** Fleisch., Musci Fl. Buitenzorg 4: 1371. 1923. (Hookeriaceae)

Type: *C. glaucocarpa* (Schwaegr.) Fleisch.

– ***glaucocarpa*** (Schwaegr.) Fleisch., Musci Fl. Buitenzorg 4: 1372. 1923. *Hypnum glaucocarpon* Reinw. *ex* Schwaegr., Spec. Musc. Suppl. 3(1): 228a. 1828. ........ E.H.; Nepal

***Chaetomitrium*** Doz. *et* Molk., Musci Fr. ined. Archip. Indici 4: 117. 1846. (Hookeriaceae)

Type: *C. elongatum* (Doz. *et* Molk.) Doz. *et* Molk.

– ***papillifolium*** Bosch *et* Lac., Bryol. Jav. 2: 50. 171. 1862.
........ E.H.; Andaman Is.

– ***sikkimense*** Gang., Mosses E. India 6: 1501. 1977.
........ E.H. Endemic

***Chinostomum*** C. Muell., Linnaea 36: 21. 1869. (Sematophyllaceae)

Type: *C. rostratum* (Griff.) C. Muell.

– ***rostratum*** (Griff.) C. Muell., Linnaea 36: 21. 1869. *Neckera rostrata* Griff., Calcutta J. Nat. Hist. 3: 70. 1843. ........ E.H.; S.I.

***Chrysocladium*** Fleisch., Musci Fl. Buitenzorg 3: 829. 1908. (Meteoriaceae)

– ***flammeum*** (Mitt.) Fleisch., Musci Fl. Buitenzorg 3: 830. 1908. *Meteorium flammeum* Mitt.. J. Linn. Soc. Bot. Suppl.. I: 88. 1859.
........ E.H.; Nepal

– ***horridum*** Dix., J. Bombay Nat. Hist. Soc. 39: 783. 1937.
........ E.H. Endemic

– ***infuscatum*** (Mitt.) Fleisch., Musci Fl. Buitenzorg 3: 830. 1908. *Meteorium infuscatum* Mitt., J. Linn. Soc. Bot. Suppl. 1: 91. 1859.
........ E.H. Endemic

– ***phaeum*** (Mitt.) Fleisch., Musci Fl. Buitenzorg 3: 830. 1908. *Meteorium phaeum* Mitt., J. Linn. Soc. Bot. Suppl. 1: 87. 1859. ........ E.H.; Nepal

– ***retrorsum*** (Mitt.) Fleisch., Musci Fl. Buitenzorg 3: 829. 1908. *Meteorium retrorsum* Mitt., J. Linn. Soc. Bot. Suppl. 1: 90. 1859. ........ S.I.

– ***tumido-aureum*** (Ren. *et* Card.) Fleisch., Musci Fl. Buitenzorg 3: 830. 1908. *Papillaria tumido-aurea* Ren. *et* Card., Bull. Soc. R. Bot. Belg. 41(1): 1. 1905. ........ E.H. Endemic

***Cinclidotus*** P. Beauv., Mag. Enc. 5: 319. 1804. (Pottiaceae)

Type: *C. fontinaloides* (Hedw.) P.Beauv.

– ***acutifolius*** Broth., Act. Soc. Sc. Fenn. 24(2): 14. 1899. ........ W.H.; Nepal

***Cirriphyllum*** Grout, Bull. Torr. Bot. Cl. 25: 222. 1898. (Brachytheciaceae)

– ***alare*** Dix., Not. R. Bot. Gard. Edinburgh 19(25): 297.11. 1938. ........ W.H.

– ***cameratum*** (Mitt.) Broth., Nat. Pfl. ed. 2, 11: 369. 1925. *Hypnum cameratum* Mitt., J. Linn. Soc. Bot. Suppl. 1: 78. 1859. ........ W.H.; E.H.; Nepal

– ***cirrhosum*** (Schwaegr.) Grout, Bull. Torr. Bot. Cl. 25: 223. 1898. *Hypnum cirrhosum* Schwaegr. *in* Schultes, Reise Glockner 365. 1804. ........ W.H.

***Claopodium*** (Lesq. *et* Jam.) Ren. *et* Card., Rev. Bryol. 20: 16. 1893. (Thuidiaceae)

– ***assurgens*** (Sull. *et* Lesq.) Card., Bull. Soc. Bot. Geneve ser. 2, 3: 283. 1911. *Hypnum assurgens* Sull. *et* Lesq., Proc. Am. Ac. Arts. Sc. 4: 279. 1859. ........ W.H.; E.H.; S.I.

– ***nervosum*** (Harv.) Fleisch., Musci Fl. Buitenzorg 4: 1504. 1923. *Hypnum nervosum* Harv. *in* Hook., Icon. Pl. Rar. 1: 24 *f.* 3. 1836. ........ W.H.; E.H.; S.I.; Nepal

– ***pellucinerve*** (Mitt.) Best, Bryologist 3: 19. 1900. *Leskea pellucinervis* Mitt., J. Linn. Soc. Bot. Suppl. 1: 130. 1859. ........ W.H.; E.H.; S.I.

– ***prionophyllum*** (C.Muell.) Broth., Nat. Pfl. 1(3): 1009. 1908. *Hypnum prionophyllum* C. Muell., Syn. 2: 481. 1851.
........ W.H.; E.H.; S.I.; Nepal

– ***vestitum*** Dix. *et* P. Vard., Arch. Bot. Bull. mens. 1(8-9): 172. 1927.
........ S.I.

***Clastobryopsis*** Fleisch., Musci Fl. Buitenzorg 4: 1179. 1923.
(Sematophyllaceae)

– ***muelleri*** (Dix.) Tix., Rev. Bryol. Lich. 43: 413. 1977. *Aptychella muelleri* Dix., Ann. Bryol. 6: 32. 1933. ........ E.H.; Nepal

– ***planula*** (Mitt.) Fleisch., Musci Fl. Buitenzorg 4: 1180. 1923. *Stereodon planulus* Mitt., Musci Ind. Or. 111. 1859.
........ E.H.; Nepal

***Clastobryum*** Doz. *et* Molk., Musci Ined. Archip. Indici 2: 41. 1845.
(Sematophyllaceae)

Type: *C. indicum* (Doz. *et* Molk.) Doz. *et* Molk.

– ***capillaceum*** (Griff.) Broth. *in* Fleisch., Musci Fl. Buitenzorg 4: 1686. 1923. *Neckera capillacea* Griff., Calcutta J. Nat. Hist. 3: 70. 1843. ........ W.H.; E.H.; S.I.; Nepal

– ***surculare*** Dix., Bull. Torr. Bot. Cl. 51: 237. 1924.
........ E.H. Endemic

– ***wichurae*** Dix., Ann. Bryol. 6: 34. 1933. ........ E.H. Endemic

***Conostomum*** Sw. *in* Web. *et* Mohr, Naturh. Reise Schwedens 122. 1804.
(Bartramiaceae)

Type: *C. arcticum* Sw.

– ***tetragonum*** (Hedw.) Lindb., Oefv. K. Vet. Ak. Foerh. 20(7): 392. 1863. *Bryum tetragonum* (Hedw.) Brid., Musc. Rec. 2(3): 59. 1803.
........ W.H.

***Coscinodon*** Spreng., Einl. Stud. Krypt. Gew. 281. 1804. (Grimmiaceae)

Type: *C. pulvinatus* Spreng.

– ***cribrosus*** (Hedw.) Spruc., Pyren. n. 267. 1847. *Grimmia cribrosa* Hedw., Spec. Musc. 76. 1801. ........ W.H.; E.H.

***Cratoneuron*** (Sull.) Spruc., Cat. Musc. Amaz. And. 21. 1867. (Amblystegiaceae)

L. Type : *C. filicinum* (Hedw.) Spruc.

– ***commutatum*** (Hedw.) Roth, Hedwigia 38: 6. 1899. *Hypnum commutatum* Hedw., Spec. Musc. 284. 1801. ........ W.H.

var. ***falcatum*** (Brid.) Moenk., Hedwigia 50: 269. 1911. *Hypnum falcatum* Brid., Musc. Rec. 2(3). 63. 1*.f.* 6.1801. ........ W.H.

– ***filicinum*** (Hedw.) Spruc., Cat. Musc. Amaz. And. 21. 1867. *Hypnum filicinum* L. *ex* Hedw., Spec. Musc. 285. 1801. ........ W.H.; Nepal

var. ***curvicaule*** (Jur.) Moenk., Hedwigia 50: 267. 1911. *Hypnum curvicaule* Jur., Verh. Zool. Bot. Ges. Wien. 14. 103. 1864. ........ Nepal

var. ***fallax*** (Brid.) Roth, Eur. Laubm. 2: 532. 1904. *Hypnum fallax* Brid., Musc. Rec. 2(2): 66. 2 *f.* 1. 1801. ........ W.H.; Nepal

***Cryptoleptodon*** Ren. *et* Card., Bull. Soc. R. Bot. Belg. 38(1): 30. 1900. (Neckeraceae)

Type: *C. flexuosus* (Harv.) Ren. *et* Card.

– ***flexuosus*** (Harv.) Ren. *et* Card., Bull. Soc. R. Bot. Belg. 38(1): 50. 1900. *Neckera flexuosa* Harv. in Hook., Icon. Pl. Rar. 1: 21 *f.* 3. 1836. ........ W.H., EH.; Nepal

– ***pluvinii*** (Brid.) Broth., Nat. Pfl. 1(3): 838. 1906. *Pilotrichum pluvinii* Brid., Mant. Musc. 141. 1819. ........ W.H.; Nepal

– ***rigidulus*** (Mitt.) Broth., Nat. Pfl. 1(3): 838. 1906. *Neckera rigidula* Wils. *ex* Mitt., J. Linn. Soc. Bot. Suppl. 1: 120. 1859. ........ W.H.

***Ctenidium*** (Schimp.) Mitt., J. Linn. Soc. Bot. 12: 509. 1869. (Hypnaceae)

Type: *C. procerrimum* (Mol.) Lindb.

– ***lychnites*** (Mitt.) Broth., Nat. Pfl. 1(3): 1048. 1909. *Stereodon lychnites* Mitt., Musci Ind. Or. 114. 1859. ........E.H.; S.I.; Nicobar Is.

***Cyathophorella*** (Broth.) Fleisch., Musci Fl. Buitenzorg 3: 1088. 1908. (Hypopterygiaceae)

– ***adiantum*** (Griff.) Fleisch., Musci Fl. Buitenzorg 3: 1094. 1908. *Neckera adiantum* Griff., Icon. Pl. As. 2 : Pl. 85-II. 1849.
........ W.H.; E.H.; Nepal

– ***anisodon*** Dix. *et* Herz., Ann. Bryol. 12: 92. 1939.
........ E.H. Endemic

– ***burkillii*** (Dix.) Broth., Nat. Pfl. ed. 2, 11: 278. 1925. *Cyathophorum burkillii* Dix., Rec. Bot. Surv. India 6: 67. 1914.
........ E.H. Endemic

– ***hookeriana*** (Griff.) Fleisch., Musci Fl. Buitenzorg 3: 1094. 1908. *Neckera hookeriana* Griff., Notul. Pl. Asiat. 2: 464. 1849.
........ E.H.; Nepal

– ***intermedia*** (Mitt.) Broth., Nat. Pfl. ed. 2, 11: 277. 1925. *Cyathophorum intermedium* Mitt., Musci Ind. Or. 148. 1859.
........ W.H.; E.H.; Nepal

– ***tonkinensis*** (Broth. *et* Par.) Broth., Nat. Pfl. ed. 2, 11: 278. 1925. *Cyathophorum tonkinense* Broth. *et* Par., Rev. Bryol. 35: 46. 1908.
........ W.H.

***Cyathothecium*** Dix., Not. R. Bot. Gard. Edinburgh 19: 300. 1938.
(Hylocomiaceae)

Type: *C. distichaceum* Dix.

– ***distichaceum*** Dix., Not. R. Bot. Gard. Edinburgh 19: 300. 14. 1938.
........ W.H.

***Cynodontium*** Schimp., Coroll. 12. 1856. (Dicranaceae)

L. Type : *C. polycarpum* (Hedw.) Schimp.

– ***gracilescens*** (Web. *et* Mohr) Schimp., Coroll. 12. 1856. *Dicranum gracilescens* Web. *et* Mohr, Bot. Taschenb. 184. 1807.
........ E.H.; Nepal

– ***polycarpum*** (Hedw.) Schimp., Coroll. 12. 1856. *Fissidens polycarpus* Hedw., Spec. Musc. 159. 1801. ........ Nepal

***Daltonia*** Hook. *et* Tayl., Musc. Brit. 80. 1818. (Daltoniaceae)

L. Type: *D. splachnoides* (Sm.) Hook. *et* Tayl.

– ***angustifolia*** Doz. *et* Molk., Ann. Sc. Nat. Bot. ser. 3, 2: 302. 1844.
........ Nepal

– ***apiculata*** Mitt., J. Linn. Soc. Bot. Suppl. 1: 145. 1859. ........ E.H.; Nepal

– ***aristifolia*** Ren. *et* Card., Rev. Bryol. 23: 105. 1896. ........ E.H.; Nepal

– ***aristifolia*** Ren. *et.* Card., Rev. Bryol. 23: 105. 1896. ssp. ***leptophylla*** Broth. *ex* Fleisch., Musci Fl., Buitenzorg 3: 966. 1908. ........ E.H.; Nepal

– ***decolyi*** Gang., Mosses E. India 6: 1473. 1977. ........ E.H. Endemic

– ***flexifolia*** Mitt., J.Linn. Soc. Bot. Suppl. 1: 146. 1859. ........ Nepal

– ***gemmipara*** Dix., J. Bombay Nat. Hist. Soc. 39: 786. 1937. ........ E.H.; Nepal

– ***himalayensis*** Dix. *et* Herz., Ann. Bryol. 12: 92. 1939. ........ E.H. Endemic

– ***marginata*** Griff., Calcutta J. Nat. Hist. 3: 270. 1843. ........ E.H.; Nepal

– ***perlaxiretis*** Dix., J. Bombay Nat. Hist. Soc. 39: 786. 1937. ........ E.H. Endemic

– ***semitorta*** Mitt., J. Linn. Soc. Bot. Suppl. 1: 146. 1859. ........ E.H. Endemic

– ***subangustifolia*** Ren. *et* Card., Bull. Soc. R. Bot. Belg. 41(1): 84. 1905. ........ E.H. Endemic

– ***subapiculata*** Hamp. *in* Gang., Mosses E. India 6: 14. 1977. ........ E.H. Endemic

***Dendrocyathophorum*** Dix., J. Bot. 74: 7. 610 f. 5. 1936. (Hypopterygiaceae)

Type: *D. assamicum* Dix.

– ***intermedium*** (Mitt.) Herz., J. Hattori Bot. Lab. 14: 65. 1955. *Cyathophorum intermedium* Mitt., J. Linn. Soc. Bot. Suppl. 1: 148. 1859. ........ E.H.; Nepal

– ***paradoxum*** (Broth.) Dix., J. Bot. 75: 126. 1937. *Hypopterygium paradoxum* Broth. in Card., Bull. Soc. Bot. Geneve ser. 2, 4: 378. 1912. ........ E.H.

***Desmatodon*** Brid., Mant. Musc. 86. 1819. (Pottiaceae)

L. Type : *D. latifolius* (Hedw.) Brid.

– ***cernuus*** (Hueb.) B.S.G., Bryol. Eur. 2: 58. 134. 1843. *Dermatodon cernuus* Hueb., Musc. Germ. 117. 1833. ........ W.H.

– ***gemmascens*** Chen, Hedwigia 80: 297. 1941. ........ W.H.; E.H.; Nepal

– ***kabir-khanii*** Broth., Mitteil. inst. Allg. Bot. Hamburg 8(2): 400. 1931. ........ W.H.

– ***leucostoma*** (R.Brown) Berggr., Musci Spetsb. n. 34. 1874. *Barbula leucostoma* R. Brown, Suppl. App. Capt. Parry's Voyage 298. 1823. ........ W.H.

***Desmotheca*** Lindb., J. Linn. Soc. Bot. 13: 184. 1872. (Orthotrichaceae)

– ***apiculata*** (Doz. *et* Molk.) Lindb. *in* Card., Ann. Jard. Bot. Buitenzorg Suppl. 1: 11. 1897. *Cryptocarpus apiculatus* Doz. *et* Molk., Ann. Sc. Nat. Bot. ser. 3, 2: 302. 1844. ........ Andaman Is.

***Diaphanodon*** Ren. *et* Card., Rev. Bryol. 22:33. 1895. (Trachypodaceae)

Type: *D. thuidioides* Ren. *et* Card.

– ***blandus*** (Harv.) Ren. *et* Card., Bull. Soc. R. Bot. Belg. 38(1): 23. 1900. *Neckera blanda* Harv. *in* Hook., Icon. Pl. Rar. 1: 22*f.* 1. 1836. ........ W.H.; E.H.; S.I.; G.P.; Nepal

– ***procumbens*** (C. Muell.) Ren. *et* Card., Bull. Soc. R. Bot. Belg. 38(1): 24. 1900. *Neckera procumbens* C. Muell., Syn. 2: 80. 1850. ........ W.H.; E.H.; S.I.; G.P.

***Dicranella*** (C. Muell.) Schimp., Coroll. 13. 1856. (Dicranaceae)

L. Type : *D. grevilleana* (Brid.) Schimp.

– ***amplexans*** (Mitt.) Jaeg., Ber. S. Gall. Naturw. Ges. 1870-71: 376. 1872. *Leptotrichum amplexans* Mitt., Musc. Ind. Or. 9. 1859. ........ Nepal

– ***divaricata*** (Mitt.) Jaeg., Ber. S. Gall. Naturw. Ges. 1870-71: 376. 1872. *Leptotrichum divaricatum* Mitt., Musc. Ind. Or. 9. 1859. ........ W.H.; E.H.; S.I.; Nepal

– ***heteromalla*** (Hedw.) Schimp., Coroll. 13. 1856. *Dicranum heteromallum* Hedw., Spec. Musc. 128. 1801. ........ W.H.; E.H.; Nepal

– ***leptoneura*** Dix., J. Bombay Nat. Hist. Soc. 39: 774. 1937.
........ E.H. Endemic

– ***macrospora*** Gang., Nov. Hedwigia 8: 145. 1964.
........ E.H. Endemic

– ***pseudosubulata*** Gang., Bull. Bot. Soc. Bengal 14: 21. 1960.
........ E.H. Endemic

– ***setifera*** (Mitt.) Jaeg., Ber. S. Gall. Naturw. Ges. 1870-71: 376. 1872. *Leptotrichum setiferum* Mitt., Musc. Ind. Or. 11. 1859. ........ E.H.

– ***spiralis*** (Mitt.) Jaeg., Ber. S. Gall. Naturw. Ges. 1877-78: 374. 1880. *Leptotrichum spirale* Mitt., Musc. Ind. Or. 158. 1859.
........ W.H.; E.H.

***Dicranodontium*** B.S.G., Bryol. Eur. 1: 159. 1847. (Dicranaceae)

Type: *D. longirostre* (Web. *et* Mohr) B.S.G.

– ***asperulum*** (Mitt.) Broth., Nat. Pfl. 1(3): 336. 1901. *Dicranum asperulum* Mitt., Musc. Ind. Or. 22. 1859. ........E.H.; Nepal

– ***attenuatum*** (Mitt.) Wils. *ex* Jaeg., Ber. S. Gall. Naturw. Ges. 1877-78: 380. 1880. *Dicranum attenuatum* Mitt., Musc. Ind. Or. 22. 1859.
........ W.H.; E.H.

– ***caespitosum*** (Mitt.) Par., Ind. Bryol. 337. 1896. *Dicranum caespitosum* Mitt., Musc. Ind. Or. 22. 1859.
........ W.H.; E.H.; Nepal

– ***capillifolium*** (Dix.) Tak., J. Hattori Bot. Lab. 31: 292. 1968. *Brothera capillifolia* Dix., J. Bombay Nat. Hist. Soc. 39: 774. 1937.
........ E.H. Endemic

– ***decipiens*** (Mitt.) Mitt. *ex* Broth., Nat. Pfl. ed.2, 10: 190. 1924. *Dicranum decipiens* Mitt., Musc. Ind. Or. 16. 1859.
........ E.H.; Nepal

– ***denudatum*** (Brid.) Britt. in Williams, N. Am. Fl. 15: 151. 1913. *Dicranum denudatum* Brid., Spec. Musc. 1: 184. 1806.
........ W.H.; E.H.; Nepal

– ***didictyon*** (Mitt.) Jaeg., Ber. S. Gall. Naturw. Ges. 1877-78: 380. 1880. *Dicranum didictyon* Mitt., Musc. Ind. Or. 21. 1859.
........ W.H.; E.H.; S.I.; Nepal

– ***didymodon*** (Griff.) Par., Ind. Bryol. 338. 1896.*Dicranum didymodon* Griff., Calcutta J. Nat. Hist. 2: 499. 1842. ........ E.H.

– ***dimorphum*** Mitt. *in* Ren. *et* Card., Bull. Soc. R. Bot. Belg. 38(1): 8. 1900. ........ E.H.; Nepal

– ***fleischerianum*** Schultze-Motel, Taxon 12: 127. 1963. ........ E.H.

– ***perviride*** Dix. *et* P. Vard., Rev. Bryol. 52: 38. 1925. ........ S.I.

– ***sordidum*** (Mitt.) Gang., Nov. Hedwigia 12: 429. 1966. *Dicranum sordidum* Wils. *ex* Mitt., Musc. Ind. Or. 18. 1859. ........ Nepal

– ***uncinatum*** (Harv.) Jaeg., Ber. S. Gall. Naturw. Ges. 1877-78: 380. 1880. *Thysanomitrium uncinatum* Harv. in Hook., J. Bot. 2: 6. 1840. ........ E.H.; Nepal

***Dicranoloma*** (Ren.) Ren., Rev. Bryol. 28(5): 85. 1901. (Dicranaceae)

– ***assimile*** (Hamp.) Par., Ind. Bryol. ed. 2,2: 24. 1904. *Dicranum assimile* Hamp., Icon. Musc. 24. 1844. ........ Nepal

– ***blumii*** (Nees) Par., Ind. Bryol. ed. 2,2: 25. 1904. *Dicranum blumii* Nees, Nov. Act. Leop.Car. 11: 131. 1823. ........ E.H.; Nepal

– ***fragile*** Broth., Nat. Pfl. ed. 2,10: 209. 1924. ........ E.H.; S.I.; Nepal

– ***gymnostomum*** (Mitt.) Par., Ind. Bryol. ed. 2,2: 27. 1904. *Dicranum gymnostomum* Mitt., J. Linn. Soc. Bot. Suppl. 1: 14. 1859. ........ E.H.

– ***subreflexifolium*** (C.Muell.) Par., Ind. Bryol. ed. 2,2: 30. 1904. *Dicranum subreflexifolium* C.Muell., Bot. Zeit. 22: 349. 1864. ........ E.H.; Nepal

***Dicranoweisia*** Lindb. *ex* Mild., Bryol. Siles. 48. 1869. (Dicranaceae)

L. Type : *D. crispula* (Hedw.) Mild.

– ***alpina*** (Mitt.) Par., Ind. Bryol. 339. 1896. *Holomitrium alpinum* Mitt., Musc. Ind. Or. 25. 1859. ........ E.H.; Nepal

– ***brevifolia*** Dix. *in* Herz., Reprium nov. Spec. Regni veg. No. 108(1): 6. 1938. ........ I.O.

– ***cirrata*** (Hedw.) Lindb. *in* Mild., Bryol. Siles. 49. 1869. *Weisia cirrata* Hedw., Spec. Musc. 69.12 f. 7-12. 1801. ........ W.H.

– ***crispula*** (Hedw.) Mild., Bryol. Siles. 49. 1869. *Weisia crispula* Hedw., Spec. Musc. 68. 12 f. 1-6. 1801. ........ W.H.

– ***indica*** (Wils.) Par., Ind. Bryol. 341. 1896. *Weisia indica* Wils., Kew J. Bot. 9: 292. 1857. ........ W.H.; E.H.; Nepal

***Dicranum*** Hedw., Spec. Musc. 126. 1801. (Dicranaceae)

L. Type : *D. scoparium* Hedw.

– ***assamicum*** Dix., J. Bombay Nat. Hist. Soc. 39: 774. 1937. ........ E.H. Endemic

– ***bonjeanii*** De Not. *in* Lisa, Elenco Musch Torino 29. 1837. ........ W.H.

– ***crispifolium*** C.Muell., Bot. Zeit. 22: 349. 1864. ........ E.H.; S.I.; Nepal

– ***dilatinerve*** Card. *et* P. Vard., Rev. Bryol. 49(3): 35, 3, a-i. 1922. ........ S.I.

– ***gymnostomum*** Mitt., J. Linn. Soc. Bot. Suppl. 1: 14. 1859. ........ E.H.; Nepal

– ***himalayanum*** Mitt., J. Linn. Soc. Bot. Suppl. 1: 14. 1859. ........ W.H.; E.H.; Nepal

– ***kashmirense*** Broth., Act. Soc. Sc. Fenn. 24(2): 9. 1890. ........ W.H.; E.H.; Nepal

– ***lorifolium*** Mitt., J. Linn. Soc. Bot. Suppl. 1.15. 1859. ........ W.H.; E.H.; Nepal

– ***majus*** Turn., Musc. Hib. 59. 1804. ........ Nepal

– ***orthophylloides*** Dix., Not. R. Bot. Gard. Edinburgh 19: 280. 1938. ........ Nepal

– ***scoparium*** Hedw., Spec. Musc. 126. 1801. *Bryum scoparium* L., 1753 nom. inval. ........ W.H.; E.H.; Nepal

– ***spurium*** Hedw., Spec. Musc. 141. 1801. ........ W.H.; E.H.

– ***undulatum*** Schrad. *ex* Brid., J.f. Bot. 1800(2): 294. 1801. ........ W.H.; E.H.

***Didymodon*** Hedw., Spec. Musc. 104. 1801. (Pottiaceae)

L. Type : *D. rigidulus* Hedw.

– ***asperifolius*** (Mitt.) Crum, Steere & Anders., Bryologist 67: 163. 1964. *Barbula asperifolia* Mitt., J. Linn. Soc. Bot. Suppl. 1: 34. 1859. ........ W.H.; E.H.

– ***canaliculatus*** Dix., Ann. Bryol. 3: 55. 1930. ........ W.H.

– ***constrictus*** (Mitt.) Saito, J. Hattori Bot. Lab. 39: 514. 1975. *Barbula constricta* Mitt., J. Linn. Soc.Bot. Suppl. 1: 33. 1859. ........ W.H.; E.H.; Nepal

– ***fallax*** (Hedw.) Zand., Phytologia 41: 28. 1978. *Barbula fallax* Hedw., Spec. Musc. 120. 1801. ........W.H.

– ***hastatus*** (Mitt.) Zand., Bull. Buffalo Soc. Nat. Sc. NY 32: 162. 1993. *Barbula hastata* Mitt., J. Linn. Soc. Bot. Suppl. 1: 34. 1859. .......W.H.; E.H.

– ***leskeoides*** Saito, J. Hattori Bot. Lab. 39: 508. 1975. ........ W.H.

– ***maschalogena*** (Ren. *et* Card.) Broth., Nat. Pfl. 1(3): 1192. 1909. *Barbula maschalogena* Ren. *et* Card., Bull. Soc. R. Bot. Belg. 41(1): 53. 1905. ........ E.H. Endemic

– ***michiganensis*** (Steere) Saito, J. Hattori Bot. Lab. 39: 517. 1975. *Barbula michiganensis* Steere in Grout, Moss. Fl.N. Am. 1: 180, f.89. 1938. ........ E.H.

– ***mittenii*** Gang., Nov. Hedwigia 8: 149. 1964. ........ W.H.; E.H.

– ***nigrescens*** (Mitt.) Saito, J. Hattori Bot. Lab. 39: 510. 1975. *Barbula nigrescens* Mitt., J. Linn. Soc. Bot. Suppl. 1: 36. 1859. ........ W.H.; E.H.

– ***rigidulus*** Hedw., Spec. Musc. 104. 1801. ........ W.H.; E.H.; S.I.

var. ***gracilis*** (Schleich. *ex* Hook. *et* Grev.) Zand., Crypt. Bryol. Lich. 2: 393. 1981 (1982). ........ W.H.

– ***stewartii*** (Bartr.) Zand., Bull. Buffalo Soc. Nat. Sc. NY 32: 162. 1993. *Barbula stewartii* Bartr., Bull. Torr. Bot. Cl. 82: 23. 1955. ........W.H.

– ***tophaceus*** (Brid.) Lisa, Elenco Muschi Torino 31. 1837. *Trichostomum tophaceum* Brid., Mant. Musc. 84. 1819. ........ W.H.

– ***vinealis*** (Brid.) Zand., Phytologia 41: 25. 1978. *Barbula vinealis* Brid., Bryol. Univ. 1: 830. 1827. ........ W.H.; Nepal

***Diphyscium*** Mohr, Obs. Bot. 34. 1803. *et* Mohr, Ind. Mus. Pl. Crypt. 3. 1903. (Buxbaumiaceae)

Type: *D. foliosum* (Hedw.) Mohr

– ***fasciculatum*** Mitt., J. Linn. Soc. Bot. Suppl. 1: 150. 1859. ........ S.I.

– ***involutum*** Mitt., J. Linn. Soc. Bot. Suppl. 1: 149. 1859. ........ E.H.; S.I.

– ***longifolium*** Griff., Calcutta J. Nat. Hist. 2: 477. 1842. ........ E.H.

***Distichium*** B.S.G., Bryol. Eur. 2: 153. 1846. (Ditrichaceae)

L.Type : *D. capillaceum* (Hedw.) B.S.G.

– ***capillaceum*** (Hedw.) B.S.G., Bryol. Eur. 2: 156. 193. 1846. *Cynodontium capillaceum* Hedw. Spec. Musc. 57. 1801. ........ W.H.; E.H.; Nepal

– ***inclinatum*** (Hedw.) B.S.G., Bryol. Eur. 2: 157. 194. 1846. *Cynodontium inclinatum* Hedw., Spec. Musc. 58. 1801. ........ W.H.; E.H.

***Distichophyllum*** Doz. *et* Molk., Musci Fr. ined. Archip. Indici 4: 99. 1846. (Hookeriaceae)

– ***decolyi*** Gang., Mosses E. India 6: 1488. 1977. ........ E.H. Endemic

– ***griffithii*** (Mitt.) Par., Ind. Bryol. 389. 1896. *Mniadelphus griffithii* Mitt., Musci Ind. Or. 145. 1859. ........ E.H.; Nepal

– ***heterophyllum*** (Mitt.) Par., Ind. Bryol. 389. 1896. *Mniadelphus heterophyllus* Wils. *ex* Mitt., Musci Ind. Or. 144. 1859. ........ E.H. Endemic

– ***humifusum*** (Mitt.) Par., Ind. Bryol. 389. 1896. *Mniadelphus humifusum* Wils. *ex* Mitt.. J. Linn. Soc. Bot. Suppl. 1: 144. 1859. .......W.H.; E.H.

– ***madurense*** Ther. *et* P. Vard., Rev. Bryol. 52: 40. 1925. ........ S.I.

– ***mittenii*** Bosch *et* Lac., Bryol. Jav. 2: 25. 149. 1861. ........ C.I.

– ***montagneanum*** (C.Muell.) Bosch *et* Lac., Bryol. Jav. 2: 23. 151. 1861. *Mniadelphus montagneanus* C. Muell., Syn. 2: 22. 1850. ........ S.I.

– ***nigricaule*** Mitt. *ex* Bosch *et* Lac., Bryol. Jav. 2: 19. 47. 1861. ........ Nepal

– ***obovatum*** (Griff.) Par., Ind. Bryol. 390. 1896. *Hookeria obovata* Griff., Calcutta J. Nat. Hist. 3: 278. 1843. ........ E.H. Endemic

– ***schmidtii*** Broth., Bot. Tidskar. 24: 122. 1901. ........ W.H.; S.I.

– ***sinuosulum*** Dix., J. Siam Soc. Nat. Hist. Suppl. 10(1): 15. 1935. ........ Nepal

– ***succulentum*** (Mitt.) Broth., Nat. Pfl. 1(3): 929. 1907. *Mniadelphus succulentus* Mitt., J. Linn. Soc. Bot. Suppl. 1: 145. 1859. ........ S.I.

***Ditrichopsis*** Broth., Sitzungsber. Ak. Wiss. Wien Math. Nat. Kl. 1.133: 560. 1924. (Ditrichaceae)

Type: *D. gymnostoma* Broth.

– ***clausa*** Broth., Symb. Sin. 4: 13. 1929. ........ E.H.

***Ditrichum*** Hamp., Flora 50: 181. 1867. (Ditrichaceae)

Type: *D. homomallum* (Hedw.) Hamp.

– ***amoenum*** (Thwait. *et* Mitt.) Par., Ind. Bryol. 391. 1896. *Cynontodium amoenum* Thwait. *et* Mitt., J. Linn. Soc. Bot. 13: 296. 1873. ........ S.I.

– ***apophysatum*** Hamp. *in* Jaeg. *ex* Gang., Bull. Bot. Soc. Bengal 13: 4. 1959. *Ditrichum apophysatum* Hamp. *in* Jaeg., Ber. S. Gall. Naturw. Ges. 1877-78: 398. 1880. ........ E.H. Endemic

– ***darjeelingense*** Ren. *et* Card., Bull. Soc. R. Bot. Belg. 41(1): 51. 1905. *Fissidens darjeelingensis* (Ren. *et* Card.) Syndow, Bot. Jahresber. 33: 61. 1906. ........ E.H. Endemic

– ***difficile*** (Dub.) Fleisch., Musci Fl. Buitenzorg 1: 300. 50. 1904. *Trichostomum difficile* Dub. in Moritzi, Syst. Verz. Zoll. Pfl. 134. 1846. .......W.H.; S.I.

– ***flexicaule*** (Schwaegr.) Hamp., Flora 50: 182. 1867. *Cynodontium flexicaule* Schwaegr., Spec. Musc. Suppl. 1(1): 113. 1811. ........ W.H.; S.I.; Nepal

– ***heteromallum*** (Hedw.) Britt., N. Am. Fl. 15: 64. 1913. *Weisia heteromalla* Hedw., Spec. Musc. 71. 1801. ........ W.H.; E.H.

– ***homomallum*** (Hedw.) Hamp., Flora 50: 182. 1867. *Didymodon homomallus* Hedw. Spec. Musc. 105.23 f.f. 1-7. 1801. ........ W.H.; E.H.

– ***laxissimum*** (Mitt.) O. Kuntze, Rev. Gen. Pl. 2: 835. 1891. *Leptotrichum laxissimum* Mitt., Musc. Ind. Or. 10. 1859. ........ E.H.; S.I. Endemic

– ***longicrure*** Ren. *et* Card., Bull. Soc. R. Bot. Belg. 38(1): 10. 1900. ........ E.H.

– ***pusillum*** (Hedw.) Hamp., Flora 50: 182. 1867. *Didymodon pusillus* Hedw., Spec. Musc. 104. 1801. ........ W.H.; E.H.; Nepal

– ***tortipes*** (Mitt.) O. Kuntze, Rev. Gen. Pl. 2: 835. 1891. *Leptotrichum tortipes* Mitt., Musc. Ind. Or. 10. 1859. ........ W.H.; E.H.; S.I.; Endemic

– ***tortuloides*** Grout, Bryologist 30: 4. 1927. ........ W.H.; E.H.

***Dixonia*** Hor. *et* Ando, Nat. & Wild Life S.E. Asia 3: 2-44. 1964. (Lembophyllaceae)

– ***thamnioides*** (Brodl. *et* Dix.) Hor. *et* Ando, Nat. & Wild Life S.E. Asia 3: 2-44. 1964.*Camptochaete thamnioides* Broth. *et* Dix., J. Bot. 53: 290. 1915. ........ Nepal

***Drepanocladus*** (C. Muell.) G. Roth, Hedwigia 38: 6. 1899. (Amblystegiaceae)

L. Type : *D. aduncus* (Hedw.) Warnst.

– ***aduncus*** (Hedw.) Warnst., Beih. Bot. Centralbl. 13: 400. 1903. *Hypnum aduncum* Hedw., Spec. Musc. 295. 1801. ........ W.H.; E.H.

– ***exannulatus*** (B.S.G.) Warnst., Beih. Bot. Centralbl. 13: 405. 1903. *Hypnum exannulatum* B.S.G., Bryol. Eur. 6: 110. 603. 1854. ........ W.H.; Nepal

– ***fluitans*** (Hedw.) Warnst., Beih. Bot. Centralbl. 13: 404. 1903. *Hypnum fluitans* Hedw., Spec. Musc. 296. 1801. ........ W.H.; Nepal

– ***uncinatus*** (Hedw.) Warnst., Beih. Bot. Centralbl. 13. 402. 417. 1903. *Hypnum uncinatum* Hedw., Spec. Musc. 289. 1801. ........ W.H.; E.H.; P.; Nepal

***Drummondia*** Hook. *in* Drum., Musci Bor. Am. n. 62. 1828. (Orthotrichaceae)

– ***duthiei*** Mitt. *ex* C. Muell., Nuov. Giorn. Bot. Ital. n. ser. 3: 106. 1896. ........ W.H.

– ***stricta*** (Mitt.) C. Muell., Nuov. Giorn. Bot. Ital. n. ser. 3: 106. 1896. *Zygodon strictus* Mitt., J. Linn. Soc. Bot. Suppl. 1: 47. 1859.
........ W.H.; E.H.; Andaman Is.

– ***thomsonii*** Mitt., J. Linn. Soc. Bot. Suppl. 1: 46. 1859. ........ W.H.

***Duthiella*** C. Muell. *in* Broth., Nat. Pfl. 1(3): 1009. 1908. (Trachypodaceae)

L. Type : *D. wallichii* (Hook.) C. Muell.

– ***declinata*** (Mitt.) Zant., Blumea 9(2): 559. 1959. *Trachypus declinatus* Mitt., J. Linn. Soc. Bot. Suppl. 1: 129. 1859.
........ W.H.; Nepal

– ***flaccida*** (Card.) Broth., Nat. Pfl. 1(3): 1010. 1908. *Trachypus flaccida* Card., Beih. Bot. Centralbl. 19(2): 117. 1905.
........ W.H.; Nepal

– ***formosana*** Nog., Trans. nat. Hist. Soc. Formosa 24: 469. 1. 1936.
........ E.H.; Nepal

– ***wallichii*** (Mitt.) C. Muell. *in* Broth., Nat. Pfl. 1(3): 1010.733. 1908. *Leskea wallichii* Mitt., J. Linn. Soc. Bot. Suppl. 1: 132. 1859.
........ W.H.; E.H.; Nepal

***Ectropothecium*** Mitt., J. Linn. Soc. Bot. 10: 180. 1868. (Hypnaceae)

– ***buitenzorgii*** (Bel.) Mitt., J. Linn. Soc. Bot. 10: 180. 1868. *Hypnum buitenzorgii* Bel. *in* Voyag., Ind. Oc. Bot. 2(Crypt.) 2: 94. 1846.
........ E.H.; Nicobar Is.

– ***compressifolium*** (Mitt.) Jaeg., Ber. S. Gall. Naturw. Ges. 1877-78: 260. 1880. *Stereodon compressifolius* Mitt., Musci Ind. Or. 99. 1859.
........ E.H.; Nepal

– ***cygnicollum*** (Mitt.) Jaeg., Ber. S. Gall. Naturw. Ges. 1877-78: 259. 1880. *Stereodon cygnicollum* Mitt., Musci Ind. Or. 100. 1859.
........ E.H.; Nicobar Is.

– ***cyperoides*** (Hook.) Jaeg., Ber. S. Gall. Naturw. Ges. 1877-78: 259. 1880. *Hypnum cyperoides* Hook., Icon. Pl. Rar. 1: 23-5. 1836.
........ W.H.; E.H.; S.I.; Nepal

var. ***papillosum*** Card. *et* Dix., Rec. Bot. Surv. India 6: 71. 1914.
........ W.H.; E.H.; S.I.; Nepal

– ***dealbatum*** (Reinw. *et* Hornsch.) Jaeg., Ber. S.Gall. Naturw. Ges. 1877-78: 264. 1880. *Hypnum dealbatum* Reinw. *et* Hornsch., Nov. Act. Ac. Car. Leop. Caes. 14(2): 729. 1829. ........ E.H.

– ***kerstanii*** Dix. *et* Herz., Ann. Bryol. 12: 96. 1939. ........ E.II. Endemic

– ***manii*** Broth., Nat. Pfl. 1(3): 1066. 1908. ........ S.I.; Andaman Is. Endemic

– ***monumentorum*** (Dub.) Jaeg., Ber. S. Gall. Naturw. Ges. 1877-78: 259. 1880. *Hypnum monumentorum* Dub. *in* Moritzi, Syst. Verz. Zoll. Pfl. 132. 1846. ........ E.H.

– ***ramuligerum*** Dix., J. Bombay Nat. Hist. Soc. 39: 793. 1937. ........ E.H. Endemic

– ***rostellatum*** (Mitt.) Jaeg., Ber. S.Gall. Naturw. Ges. 1877-78: 258. 1880. *Stereodon rostellatus* Mitt., Musci Ind. Or. 100. 1859. ........ E.H.; Nepal

– ***sikkimense*** (Ren. *et* Card.) Ren. *et* Card., Bull. Soc. R. Bot. Belg. 41(1):109. 1905. *Hypnum hamulosum* B.S.G., var. : ***sikkimense*** Ren. *et* Card., Bull. Soc. R. Bot. Belg. 34(2): 78. 1896. ........W.H.; E.H.

***Encalypta*** Hedw., Spec. Musc. 60. 1801. (Encalyptaceae)

Type: *E. alpina* Sm.

– ***alpina*** Sm., Engl. Bot. 20: 1419. 1805. ........ W.H.

– ***ciliata*** Hedw., Spec. Musc. 61. 1801. ........ W.H.

– ***rhabdocarpa*** Schwaegr., Spec. Musc. Suppl. 1(1): 56. 16. 1811. ........ W.H.

– ***streptocarpa*** Hedw., Spec. Musc. 62. 10 f. 10-15. 1801. ........ W.H.

– ***tibetana*** Mitt., J. Linn. Soc. Bot. Suppl. 1: 42. 1859. .......W.H.

– ***vulgaris*** Hedw., Spec. Musc. 60. 1801. ........ W.H.

***Endotrichella*** C. Muell., Linnaea 37: 156. 1872. (Pterobryaceae)

L. Type : *E. dictrichae* C. Muell.

– ***elegans*** (Doz. *et* Molk.) Fleisch. in Broth., Nat. Pfl. 1(3): 782. 585. 1906. *Endotrichum elegans* Doz. *et* Molk., Ann. Sc. Nat. Bot. ser. 3, 2: 303. 1844. ........ E.H.

***Entodon*** C.Muell., Linnaea 18: 704. 1845. (Entodontaceae)

L. Type : *E. cladorrhizans* (Hedw.) C. Muell.

– ***chloropus*** Ren. *et* Card., Bull. Soc. R. Bot. Belg. 38(1): 34. 1900. ........ E.H.; Nepal

– ***concinnus*** (De Not.) Par., Ind. Bryol. ed. 2, 2: 130. 1904. *Hypnum concinnum* De Not., Mem. R. Ac. Sc. Torino 39: 220. 1836. ........ I.O.

– ***curvatus*** (Griff.) Jaeg., Ber. S. Gall. Naturw. Ges. 1876-77: 293. 1878. *Neckera curvata* Griff., Calcutta J. Nat. Hist. 3: 65. 1843. ........ E.H.; S.I. Endemic

– ***flavescens*** (Hook.) Jaeg., Ber. S. Gall. Naturw. Ges. 1876-77: 293. 1878. *Neckera flavescens* Hook., Trans. Linn. Soc. London 9: 314. 1808. ........ W.H.; Nepal

– ***laetus*** (Griff.) Jaeg., Ber. S. Gall. Naturw. Ges. 1876-77: 295. 1878. *Neckera laeta* Griff., Calcutta J. Nat. Hist. 3 :67. 1843. ........ E.H.; S.I.; Nepal

– ***luridus*** (Griff.) Jaeg., Ber. S. Gall. Naturw. Ges. 1876-77: 294. 1878. *Neckera lurida* Griff., Calcutta J. Nat. Hist. 3: 66. 1843. ........ E.H.; Nepal

– ***luteonitens*** Ren. *et* Card., Bull. Soc. R. Bot. Belg. 38(1): 34. 1900. ........ E.H. Endemic

– ***macropodus*** (Hedw.) C. Muell., Linnaea 18: 707. 1845. *Neckera macropoda* Hedw., Spec. Musc. 207. 1801. ........ E.H.; Nepal

– ***myurus*** (Hook.) Hamp., Linnaea 20: 82. 1847. *Pterogonium myurum* Hook., Musci Exot. 2: 148. 1819. ........ W.H.; E.H.; Nepal

– ***nepalensis*** Mizushima, Hara, Fl. E. Him. 584. f.42. 1966. ........ Nepal

– ***ovicarpus*** Dix., J. Bombay Nat. Hist. Soc. 39: 789. 1937. ........ E.H. Endemic

– ***plicatus*** C.Muell., Linnaea 18: 706. 1845. ........ W.H.; E.H.; C.I.; G.P.; S.I.; Nepal

– ***prorepens*** (Mitt.) Jaeg., Ber. S. Gall. Naturw. Ges. 1876-77: 294. 1878. *Stereodon prorepens* Mitt., Musci Ind. Or. 107. 1859. ........ W.H.; E.H.; S.I.; Nepal

– ***pulchellus*** (Griff.) Jaeg., Ber. S. Gall. Naturw. Ges. 1876-77: 294. 1878. *Neckera pulchella* Griff., Calcutta J. Nat. Hist. 3: 66. 1843. ........ E.H. Endemic

– ***rubicundus*** (Mitt.) Jaeg., Ber. S. Gall. Naturw. Ges. 1876-77: 285. 1878. *Stereodon rubicundus* Mitt., Musci Ind. Or. 108. 1859. ........ W.H.; E.H.; S.I.; Andaman Is.; Nepal

– ***scariosus*** Ren. *et* Card., Bull. Soc. R. Bot. Belg. 34(2): 75. 1896. ........ E.H. Endemic

– ***subplicatus*** Ren. *et* Card., Bull. Soc. R. Bot. Belg. 38(1): 35. 1900. ........ W.H.; E.H. Endemic

***Entosthodon*** Schwaegr., Spec. Musc. Suppl. 2(1): 44. 1823. (Funariaceae)

Type: *E. templetonii* (Sm.) Schwaegr.

– ***buseanus*** Doz. *et* Molk., Bryol., Jav. 1: 31.22 f. 1-23. 1855. ........ S.I.

– ***diversinervis*** C.Muell., Bot. Zeit. 11: 20. 1853. ........ S.I.

– ***nutans*** Mitt., J. Linn. Soc. Bot. Suppl. 1: 55. 1859. ........ W.H.; S.I.; C.I.; P.; R.; G.P.

– ***perrottetii*** (Mont.) C. Muell., Bot. Zeit. 11: 19. 1853. ........ S.I.

– ***physcomitrioides*** Mont., Ann. Sc. Nat. Bot. ser. 2, 17: 253. 1842. ........ S.I.

– ***pilifer*** Mitt., J. Linn. Soc. Bot. Suppl. 1: 55. 1859. ........ W.H.

– ***planifolius*** Thwait. *et* Mitt., J. Linn. Soc. Bot. 13: 304. 1873. ........ S.I.

– ***pulchra*** Dix. *et* P. Vard., Arch. Bot. 1(8-9): 169.5 f. 3. 1927. ........ S.I.

– ***submarginatus*** C.Muell., Bot. Zeit. 11: 20. 1853. ........ S.I.

– ***wallichii*** Mitt., J. Linn. Soc. Bot. Suppl. 1: 55. 1859. *Funaria wallichii* (Mitt.) Broth., Nat, Pfl. 1(3): 525. 1903. ........ W.H.; E.H.; Nepal

– ***wichurae*** Fleisch., Musci Fl. Buitenzorg 2: 481. 1904. *Funaria wichurae* (Fleisch.) Broth., Nat. Pfl. 1(3): 1204. 1909. ........ W.H.; E.H.

***Epipterygium*** Lindb., Oefv. Svensk. Vet. Ak. Foerh. 19: 603. 1863. (Bryaceae)

L. Type : *E. wrightii* (Sull.) Lindb.

– ***tozeri*** (Grev.) Lindb., Oefv. Svensk. Vet. Ak. Foerh. 21: 576. 1865. *Bryum tozeri* Grev., Scott. Crypt. Fl. 5: 285. 1827. ........ W.H.; E.H.

***Eriopus*** C. Muell., Bot. Zeit. 5: 828. 1847. (Hookeriaceae)

– ***remotifolius*** C.Muell., Bot. Zeit. 5: 828. 1847. ........ E.H.

– ***spinosus*** Nog., J.Sc. Hiroshima Univ. B(2) Bot. 3: 51. 1937. ........ Nepal

***Erpodium*** (Brid.) Brid.,Bryol. Univ. 2: 788. 1827. (Erpodiaceae)

Type : *E. domingense* (Spreng.) Brid.

– ***mangiferae*** C.Muell., Linnaea 37: 178. 1873. ........ W.H.; E.H.; G.P.; C.I.; S.I.; Nepal

***Erythrodontium*** Hamp., Vid. Medd. Naturh. For. Kjoebenh. ser. 3, 2: 279. 1870. (Entodontaceae)

Type: *E. warmingii* Hamp.

– ***julaceum*** (Schwaegr.) Par., Ind. Bryol. 436. 1896. *Neckera julacea* Hook. *ex* Schwaegr, Spec. Musc. Suppl. 3(1): 245. 1828. ........ W.H.; E.H.; C.I.; S.I.; Nepal

***Eucladium*** B.S.G., Bryol. Eur. 1: 93. 1846. (Pottiaceae)

Type: *E. verticillatum* (Brid.) B.S.G.

– ***verticillatum*** (Brid.) B.S.G., Bryol. Eur. 1: 9. 40. 1846. *Weisia verticillata* Brid., J.f. Bot. 1800(2): 283. 1801. ........ W.H.

***Eurhynchium*** B.S.G., Bryol. Eur. 5: 217. 1854. (Brachytheciaceae)

– ***dumosum*** (Mitt.), Jaeg., Ber. S. Gall. Naturw. Ges. 1876-77: 346. 1878. *Hypnum dumosum* Mitt., J. Linn. Soc. Bot. Suppl. 1: 80. 1859. ........ E.H.; Nepal

– ***hians*** (Hedw.), Lac., Ann. Mus. Bat. Lugd. Bot. 2: 299. 1866. *Hypnum hians* Hedw., Spec. Musc. 272. 1801. ........ W.H.; E.H.; Nepal

– ***muelleri*** (Jaeg.) Bartr., Bishop Mus. Bull. 101: 214. 1933. *Rhynchostegium muelleri* Jaeg., Ber. S. Gall. Naturw. Ges. 1876-77: 378. 1878. ........ W.H.; E.H.; Nepal

var. ***minus*** Dix., J. Bombay Nat. Hist. Soc. 39: 788. 1937. ........ W.H.; E.H.; Nepal

– ***ovatum*** Card. *in* P. Vard., Rev. Bryol. 51: 11. 1924. ........ S.I.

– ***praelongum*** (Hedw.) B.S.G., Bryol. Eur. 5: 224. 524. 1854. *Hypnum praelongum* L. *ex* Hedw., Spec. Musc. 258. 1801. ........ W.H.; E.H.

– ***pulchellum*** (Hedw.) Jenn., Man. Moss. West Pennsylv. 350. 1913. *Hypnum pulchellum* Brid. *ex* Hedw., Spec. Musc. 256. 68 f. 1-4. 1801. ........ W.H.

– ***riparioides*** (Hedw.) Richs., Ann. Bryol. 9:135. 1937. *Hypnum riparioides* Hedw., Spec. Musc. 242. 1801. ........ W.H.; E.H.; S.I.; Nepal

– ***striatum*** (Hedw.) Schimp., Coroll. 119. 1856. *Hypnum striatum* Hedw., Spec. Musc. 275. 1801. ........ W.H.

– ***swartzii*** (Tum.) Curnow, Rabenh. Bryoth. Eur. 12: 593. 1862. *Hypnum swartzii* Turn., Musc. Hib. 151. 1804. ........ E.H.; Nepal

***Exodictyon*** Card., Rev. Bryol. 26: 6. 1899. (Calymperaceae)

– ***blumei*** (C.Muell.) Fleisch., Musci Arch. Indici n. 58. 1899. *Syrrhopodon blumei* Nees *ex* Hamp., Bot. Zeit. 5: 924. 1847. ........ E.H.; Nicobar Is.

***Fabronia*** Raddi, Atti Acc. Sc. Siena 9: 230. 1808. (Fabroniaceae)

Type: *F. pusilla* Raddi

– ***assamica*** Dix., J. Bombay Nat. Hist. Soc. 39: 787. 1937. ........ E.H. Endemic

– ***ciliaris*** (Brid.) Brid., Bryol. Univ. 2: 171. 1827. *Hypnum ciliare* Brid., Musc. Rec. Suppl. 2: 155. 1812. ........ W.H.

– ***curvirostris*** Doz. *et* Molk., Ann. Sc. Nat. Bot. ser. 3, 2: 304. 1844. ........ S.I.

– ***goughii*** Mitt., J. Linn. Soc. Bot. Suppl. 1: 75. 1859. ........ W.H.; S.I.

– ***madurensis*** Dix. *et* Vard., Arch. Bot. 1(8-9): 171.3 f.8. 1927. ........ W.H.; S.I.

– ***minuta*** Mitt., J. Linn. Soc. Bot. Suppl. 1: 76. 1859. ........ W.H.; C.I.; Endemic

– ***pusilla*** Raddi, Atti Acc. Sc. Siena 9: 231. 1. 1808. ........ I.O.

– ***schensiana*** C.Muell., Nuov. Giorn. Bot. Ital. n.ser. 4: 262. 1897. ........ Nepal

– ***schmidii*** C.Muell., Bot. Zeit. 12: 558. 1854. ........ S.I.

– ***secunda*** Mont., Ann. Sc. Nat. Bot. ser. 2, 17: 251. 1842. ........ E.H.; S.I.; Nepal

***Fissidens*** Hedw., Spec. Musc. 152. 1801. (Fissidentaceae)

L. Type : *F. bryoides* Hedw.

– ***alanii*** Gang., Mosses E. India 3: 567. 1972. ........ E.H. Endemic

– ***anomalus*** Mont., Ann. Sc. Nat. Bot. ser. 2, 17: 252. 1842. ........ W.H.; E.H.; S.I.; Nepal

– ***areolatus*** Griff., Calcutta J. Nat. Hist. 2: 506. 1842. ........ E.H.; Nepal

– ***arnigadhensis*** Broth. in Bruehl *ex* Kumar, Miscell. Bryol. Lich. 8(6): 118. 1979. ........ W.H.

– ***arunii*** Srivastava *et* Norkett, Proc. Bot. Soc. Kanpur (India) 1: 55. 1988. ........ W.H.; C.I.

– ***asperisetus*** Lac., Naturk. Verh. K. Ak. Wet. Amsterdam 13: 2. 1872. ........ S.I.; Andaman Is.

var. ***andamanensis*** Gang., Mosses E. India 3: 507. 1972. ........ S.I.; Andaman Is.

– ***biformis*** Mitt., J. Linn. Soc. Bot. Suppl. 1: 141. 1859. *F. coorgensis* Broth., Rec. Bot. Surv. India 1: 316. 1899. ........ S.I.; Nepal

– ***bilaspurense*** Gang., Bull. Bot. Soc. Bengal 11: 66. 5. 1957. ........ C.I.; G.P. Endemic

– ***bryoides*** Hedw., Spec. Musc. 153. 1801. ........ W.H.; E.H.; S.I.; G.P.; Nepal

– ***ceylonensis*** Doz. *et* Molk., Ann. Sc. Nat. Bot. ser. 3, 2: 304. 1844. ........ W.H.; E.H.; S.I.; G.P.; Nepal

– ***crenulatus*** Mitt., J. Linn. Soc. Bot. Suppl. 1: 140. 1859.

var. ***crenulatus*** ........ C.I.; G.P.; Nepal

var. ***titalyanus*** (C.Muell.) Gang., Mosses E. India 3: 506. 1972. ........ C.I.; G.P.; Nepal

– ***cristatus*** Wils. *ex* Mitt., J. Linn. Soc. Bot. Suppl. 1: 137. 1859. ........ W.H.; E.H.; S.I.; Nepal

– ***curvato-involutus*** Dix., Not. R. Bot. Gard. Edinburgh 19: 279. 1938. .......W.H.; E.H.; C.I.; G.P.; R.

– ***curvatoxiphioides*** Dix. *et* P. Vard., Arch. Bot. 1(8-9): 163.3 f. 1. 1927. ........ W.H.; S.I.; Nepal

– ***diversifolius*** Mitt., J. Linn. Soc. Bot. Suppl. 1: 140. 1859. ........ W.H.; E.H.; S.I.; C.I.; G.P.; Nepal

var. ***rubricaulis*** (Dix.) Norkett *in* Gang., Mosses E. India 3: 494. 1972. *F. rubricaulis* Dix., J. Bombay Nat. Hist. Soc. 39: 772. 1937. ........ E.H.; Nepal

– ***elongatus*** Mitt., J. Linn. Soc. Bot. Suppl. 1: 139. 1859. ........ E.H. Endemic

– ***firmus*** Mitt., J. Linn. Soc. Bot. Suppl. 1: 139. 1859. ........ S.I.

– ***ganguleei*** Norkett ex Gang., Mosses E. India 2: 527. 1971. ........ E.H.; S.I.; Nepal

– ***geminiflorus*** Doz. et Molk., Pl. Jungh. 316. 1854. ........ C.I.

– ***geppii*** Fleisch., Musci Fl. Buitenzorg 1: 26. 1904. ........ W.H.

– ***grandifrons*** Brid., Spec. Musc. 1: 170. 1806. ........ W.H.; S.I.; Nepal

– ***griffithii*** Gang., Nov. Hedwigia 8: 143. 1964. ........ S.I.

– ***hyalinus*** Hook. *et* Wils., Hook. J. Bot. 3: 89. 2. 1840. ........ W.H.

– ***incognitus*** Gang., Bull. Bot. Soc. Bengal 11: 70. 9. 1957. ........ E.H.; S.I. Endemic

– ***intromarginatulus*** Bartr., Rev. Bryol. Lich. 23: 242. 1954. ........ W.H.; E.H.; S.I.; C.I.; Nepal

– ***involutus*** Wils. *ex* Mitt., J. Linn. Soc. Bot. Suppl. 1: 138. 1859. ........ W.H.; E.H.; C.I.; Nepal

– ***javanicus*** Doz. *et* Molk., Bryol. Jav. 1: 11. 3. 1855.
........ W.H.; E.H.; Andaman Is.; Nepal

– ***jungermannioides*** Griff., Calcutta J. Nat. Hist. 2: 505. 1842.
........ E.H. Endemic

– ***kalimpongensis*** Gang., Bull. Bot. Soc. Bengal 11: 71. 1957.
........ W.H.; E.H.; S.I.; Nepal

– ***kurzii*** C.Muell., Linnaea 37: 163. 1872. ........ G.P.

– ***laxitextus*** Broth. *ex* Gang., Nov. Hedwigia 8: 144. 1964.
........ E.H.; Nepal

– ***leptopelma*** Dix., J. Bombay Nat. Hist. Soc. 39: 773. 1937.
........ E.H.; S.I. Endemic

– ***longisetus*** Griff., Calcutta J. Nat. Hist. 2: 503. 1842.
........ E.H. Endemic

– ***macrosporoides*** Dix. *et* P. Vard., Ann. Crypt. Exot. 3: 169. 1. f. 3. 1930. ........ W.H.; S.I.; C.I.

– ***minutus*** Thwait. *et* Mitt., J. Linn. Soc. Bot. 13: 323. 1873.
........ S.I.; Nepal

– ***mittenii*** Par., Ind. Bryol. 477. 1896. ........ S.I.; Nepal

– ***mussureinsis*** C.Muell. in Brühl., Rec. Bot. Surv. India 13: 123. 1931.
........ W.H.

– ***nobilis*** Griff., Calcutta J. Nat. Hist. 2: 505. 1842.
........ W.H.; E.H.; Nepal

– ***nymanii*** Fleisch., Musci Fl. Buitenzorg 1: 19.1. 1904. *F. himalayanus* Broth. in Bruehl, Rec. Bot. Surv. India 13: 123. 1931.
........ W.H.; S.I.; Nepal

– ***obscurus*** Mitt., J. Linn. Soc. Bot. Suppl. 1: 138. 1859.
........ W.H.; E.H.; C.I.; Nepal

– ***orishae*** Gang., Nov. Hedwigia 8: 140. 1964.
........ C.I.; G.P. Endemic

– ***perplexans*** Dix., Rec. Bot. Surv. India 9: 305. 1. 1926. ........ W.H.

– ***plagiochiloides*** Besch, J. de Bot. 12: 293. 1898. ........ Nepal

– ***pokharensis*** Norkett *ex* Kumar, Miscell. Bryol. Lich. 8(6): 120. 1979. ........ W.H.

– ***polysetulus*** Gang., Mosses E. India 2: 525. 1971.
........ E.H. Endemic

– ***pulchellus*** Mitt., J. Linn. Soc. Bot. Suppl. 1: 140. 1859.
........ E.H.; Nepal

– ***rambii*** Gang., Mosses E. India 2: 475. 1971. ........ E.H. Endemic

– ***ranchiensis*** Gang., Bull. Bot. Soc. Bengal 11: 68.7. 1957.
........ S.I.; C.I.; Nepal

– ***ranuii*** Gang., Nov. Hedwigia 8: 141. 1964. ........ C.I. Endemic

– ***rigidiusculus*** Gang., Mosses E. India 2: 484. 1971. *F. rigidiusculus* Broth. *in* Bruehl., Rec. Bot. Surv. India 13(1): 126. 1931.
........ E.H. Endemic

– ***robinsonii*** Broth., Philipp. J. Sc. C 13: 204. 1918.
........ Andaman Is.; Nicobar Is.

– ***semperfalcatus*** Dix., J. Siam Soc. Nat. Hist. Suppl. 10:2. 1935.
........ W.H.; E.H.; G.P.; Andaman Is.; Nepal

– ***splachnobryoides*** Broth. *in* Schum. *et* Lauterb., Fl. Deutsch. Schutzgeb. Suedsee 81. 1900. ......W.H.; E.H.; S.I.; C.I.; G.P.; Nepal

– ***subbryoides*** Gang., Nov. Hedwigia 8: 141. 1964. *F. subbryoides* Hamp., *in* ed. in shed. .......E.H.; Andaman Is.; Nepal

– ***subpalmatus*** C. Muell., Linnaea 37: 164. 1872.
........ S.I.; C.I.; G.P.; Nepal

– ***subpulchellus*** Norkett in Gang., Mosses E. India 2: 521. 1971.
........ E.H. Endemic

– ***sylvaticus*** Griff., Calcutta J. Nat. Hist. 2: 507. 1842.
........ E.H. Endemic

var. ***zippelianus*** (Doz. *et* Molk.) Gang., Mosses E. India 3: 537. 1972. *F. zippelianus* Doz. & Molk., Syst. Verzeichn. 29. 1854.
........ E.H. Endemic

– ***taxifolius*** Hedw., Spec. Musc. 155.39 f. 1-5. 1801.
........ W.H.; E.H.; S.I.; G.P.; Andaman Is.; Nepal

– ***virens*** Thwait. *et* Mitt., J. Linn Soc. Bot. 13: 324. 1873.
........ W.H.; S.I.; Nepal

– ***viridulus*** (Sw.) Wahlenb., Fl. Lapp. 334. 1812. *Dicranum viridulum* Sw., Monthl. Rev. 34: 538. 1801. ........ W.H.; Nepal

– ***walkeri*** Broth., Rec. Bot. Surv. India 1: 315. 1899. ....... S.I.

– ***xiphioides*** Fleisch., Hedwigia 38: 125. 1899.
........ W.H.; S.I.; G.P.; C.I.; Andaman Is.; Nepal

– ***zollingeri*** Mont., Ann. Sc. Nat. Bot. ser. 3, 4: 114. 1845.
........ S.I.; Andaman Is.

***Fleischerobryum*** Loesk., Stud. Morph. Syst. Laubm. 127. 1910.
(Bartramiaceae)

– ***longicolle*** (Hamp.) Loesk., Morph. Syst. Stud. Laubm. 127. 1910. *Bartramia longicollis* Hamp. *in* C. Muell., Syn. 1: 478. 1848.
........ W.H.; E.H.; Nepal

– ***macrophyllum*** Broth., Philipp. J. Sc. 31: 285. 1926. ........ E.H.

***Floribundaria*** Fleisch., Hedwigia 44: 301. 1905. (Meteoriaceae)

L. Type : *F. floribunda* (Doz. *et* Molk.) Fleisch.

– ***armata*** Broth., Symb. Sin. 4 : 83. 2 *f.* 9. 1929. ........ E.H.; Nepal

– ***aurea*** (Mitt.) Broth., Nat. Pfl. 1(3): 822. 1906. *Meteorium aureum* Mitt., J. Linn. Soc. Bot. Suppl. 1: 89. 1859. ........ E.H.; Nepal

– ***chloronema*** (C. Muell.) Broth., Nat. Pfl. 1(3): 822. 1906. *Papillaria chloronema* C. Muell. in Ren. *et* Card., Bull. Soc. R. Bot. Belg. 34(2): 69. 1896. ........ E.H.; S.I.; Nepal

– ***chrysonema*** (C.Muell.) Broth., Nat. Pfl. 1(3): 822. 1906. *Papillaria chrysonema* C. Muell. *in* Ren. *et* Card., Bull. Soc. R. Bot. Belg. 34(2): 69. 1896. ........ E.H.

– ***commutata*** (Mitt.) Broth., Nat. Pfl. 1(3): 822. 1906. *Meteorium commutatum* Mitt., J. Linn. Soc. Bot. Suppl. 1: 89. 1859. ........ E.H.

– ***floribunda*** (Doz. *et* Molk.) Fleisch., Hedwigia 44: 302. a-i. 1905. *Leskea floribunda* Doz. *et* Molk., Ann. Sc. Nat. Bot. ser. 3, 2: 310. 1844. ........ W.H.; E.H.; S.I.; Nepal

– ***leptonema*** (C. Muell.) Broth., Nat. Pfl. 1(3): 822. 1906. *Papillaria leptonema* C. Muell. *in* Ren. *et* Card., Bull. Soc. R. Bot. Belg. 34(2): 70. 1896. ........ W.H.; E.H.

– ***sparsa*** (Mitt.) Broth., Nat. Pfl. 1(3): 822. 1906. *Meteorium sparsum* Mitt., J. Linn. Soc., Bot. Suppl. 1: 158. 1859. ........ E.H.; S.I.; Nepal

– ***thuidioides*** Fleisch., Musci Fl. Buitenzorg 3: 824. 147. 1908.
........ S.I.

– ***walkeri*** (Ren. *et* Card.) Broth., Nat. Pfl. 1(3): 822. 1906. *Papillaria walkeri* Ren. *et* Card., Bull. Soc. R. Bot. Belg. 34(2): 70. 1896. ........ W.H.; E.H.; S.I.

***Foreauella*** Dix. *et* P. Vard., Arch. Bot. 1(8-9): 175. 1927. (Sematophyllaceae)

Type: *F. indica* (Schwaegr.) Dix. *et* P. Vard.

– ***orthothecia*** (Schwaegr.) Dix. *et* P. Vard., J. Bot. 75: 129. 1937. *Hypnum orthothecium* Schwaegr., Spec. Musc. Suppl. 3(1): 220b. 1827. ........ E.H.; S.I.; C.I.; Nepal

***Forsstroemia*** Lindb., Oefv. K. Vet. Ak. Foerh. 19: 605. 1863. (Leucodontaceae)

Type: *F. trichomitria* (Hedw.) Lindb.

– ***inclusa*** Card. *et* Dix., J. Bot. 50: 147. 517 f.2. 1912. ........ W.H.; E.H.

– ***indica*** (Mont.) Par., Ind. Bryol. 499. 1896. *Pterogonium indicum* Mont., Ann. Sc. Nat. Bot. ser. 2, 17: 250. 1812. ........ I.O.

– ***secunda*** Dix. *et* Badhw., Rec. Bot. Surv. India 12: 176. 1938. ........ W.H.

***Funaria*** Hedw., Spec. Musc. 172. 1801. (Funariaceae)

L. Type : *F. hygrometrica* Hedw.

– ***calcarea*** Wahlenb., Vet. Ak. Nya Handl. 27: 137.4 f.2. 1806. ........ W.H.

var. ***mediterranea*** (Lindb.) C. Jens. et Medel., Bot. Not. 1929: 47. 1929. *F. mediterranea* Lindb., Oefv. K. Vet. Ak. Foerh. 20: 399. 1863. ........ W.H.

– ***capillipes*** Broth., Act. Soc. Sc. Fenn. 24(7): 20. 1899. ........W.H.

– ***excurrentinervis*** Card. *et* P. Vard., Rev. Bryol. 50: 18.4. 1923. ........ S.I.

– ***hygrometrica*** Hedw., Spec. Musc. 172. 1801. ........ W.H., E.H., S.I.; C.I.; G.P.

var. ***calvescens*** (Schwaegr.) Mont., Ann. Sc. Nat. Bot. ser. 2, 12: 54. 1839. *F. calvescens* Schwaegr., Spec. Musc. Suppl. 1(2): 77. 1816. ........ W.H.; E.H.; S.I.; C.I.; G.P.

– ***koelzii*** Bart., Rev. Bryol. Lich. 29(3-4): 167. 1960. ........ W.H.

– ***orthocarpa*** Mitt., J. Linn. Soc. Bot. Suppl. 1: 56. 1859. ........ W.H.

– ***wijkii*** R.S. Chopra, Taxon. Indian Mosses: 185. 1975. ........ W.H.

***Gammiella*** Broth., Nat. Pfl. 1(3): 1067. 1908. (Sematophyllaceae)

Type: *G. pterogonioides* (Griff.) Broth.

– ***pterogonioides*** (Griff.) Broth., Nat. Pfl. 1(3): 1067. 763. 1908. *Pleuropus pterogonioides* Griff., Calcutta J. Nat. Hist. 3: 274. 1842. ........ E.H.

***Ganguleea*** Zand., Phytologia 65: 427. 1989. (Pottiaceae)

Type: *G. angulosa* (Broth. et Dix.) Zand.

– ***angulosa*** (Broth. *et* Dix.) Zand., Phytologia 65: 427. 1989. *Merceyopsis angulosa* Broth. *et* Dix., J. Bot. 48: 302. 1910. ........ W.H.; E.H.; Nepal

***Garckea*** C. Muell., Bot. Zeit. 3: 865. 1845. (Ditrichaceae)

Type: *G. phascoides* C. Muell.

– ***abbreviata*** Dix. *et* P. Vard., Arch. Bot. 1(8-9): 163. 1927. ........ S.I.

– ***flexuosa*** (Griff.) Marg. & Nork., J. Bryol. 7: 440. 1973. *Grimmia flexuosa* Griff., Calcutta J. Nat. Hist. 2(8): 492. 1842. ........E.H.; S.I.; G.P.; C.I.; Andaman Is.; Nepal

***Garovaglia*** Endl., Gen. Pl. 57. 1836. (Pterobryaceae)

Type: *G. plicata* (Brid.) Bosch *et* Lac.

– ***plicata*** (Brid.) Bosch *et* Lac., Bryol. Jav. 2: 79. 1863. *Esenbeckia plicata* Brid., Bryol. Univ. 2: 754. 1827. ........ E.H.

***Glossadelphus*** Fleisch., Musci Fl. Buitenzorg 4: 1351. 1923. (Sematophyllaceae)

– ***anisopterus*** (Card. *et* P. Vard.) Broth., Nat. Pfl. ed. 2, 11: 444. 1925. *Ectropothecium anisopterum* Card. *et* P. Vard., Rev. Bryol. 50: 79. 1923. ........ S.I.

– ***bilobatus*** (Dix.) Broth., Nat. Pfl. ed. 2, 11: 535. 1925. *Taxithelium bilobatum* Dix., Bull. Torr. Bot. Cl. 51: 244. 1924. ........ E.H.

– ***glossoides*** (Bosch *et* Lac.) Fleisch., Musci Fl. Buitenzorg 4: 1358. 1923. *Hypnum glossoides* Bosch *et* Lac., Bryol. Jav. 2: 146, 243. 1866. ........ Nepal

– ***isopterygioides*** (Dix.) Broth., Nat. Pfl. ed. 2, 11: 444. 1925. *Taxithelium isopterygioides* Dix., J. Bot. 53. 295. 540. f. 9. 1915. ........ S.I.

– ***ivoreanus*** (Mitt.) Fleisch., Musci Fl. Buitenzorg 4: 1352. 1923. *Stereodon ivoreanus* Mitt., J. Linn. Soc. Bot. Suppl. 1: 105. 1859. ........ S.I.; Nepal

– ***spathulifolium*** R.S. Chopra, Taxon. Indian Mosses: 521. 1975. ........ E.H.

– ***vivicolor*** (Broth. *et* Dix.) Broth., Nat. Pfl. ed. 2, 11: 444. 1925. *Taxithelium vivicolor* Broth. et Dix., Rec. Bot. Surv. India 6: 86. 1 f. 4. 1914. ........ S.I.

– ***zollingeri*** (C.Muell.) Fleisch., Musci Fl. Buitenzorg 4: 1355. 1923. *Hypnum zollingeri* C.Muell., Syn. 2: 241. 1851. ........ W.H.; E.H.; S.I.; G.P.

***Gollania*** Broth., Nat. Pfl. 1(3): 1054. 1908. (Hypnaceae)

– ***clarescens*** (Mitt.) Broth., Nat. Pfl. 1(3): 1055. 1908. *Stereodon clarescens* Mitt., Musci Ind. Or. 100. 1859. ........ W.H.; Nepal

– ***cylindricarpa*** (Mitt.) Broth., Nat. Pfl. 1(3): 1055. 1908. *Hylocomium cylindricarpum* Mitt., Trans. Linn. Soc. Bot. Lond. ser. 2, 3: 1. 1891. ........ I.O.

– ***ruginosa*** (Mitt.) Broth., Nat. Pfl. 1(3): 1055. 1908. *Hylocomium ruginosum* Mitt., Trans. Linn. Soc. Bot. London ser. 2, 3: 178. 1891. ........ W.H.

***Grimmia*** Hedw., Spec. Musc. 75. 1801. (Grimmiaceae)

L. Type: *G. plagiopodia* Hedw.

– ***anodon*** B.S.G., Bryol. Eur. 3: 110. 236. 1845. ........ W.H.

– ***apiculata*** Hornsch., Flora 2: 443. 1819. ........ W.H.

– ***apophysata*** Hamp. *ex* Gang., Nov. Hedwigia 12: 428. 1966. ........ E.H. Endemic

– ***donniana*** Sm., Engl. Bot. 18: 1259. 1804. ........ W.H.; Nepal

– ***elongata*** Kaulf. *in* Sturm, Deutschl. Fl. 2(15): 14. ic. 1816. ........ E.H.; Nepal

– ***funalis*** (Schwaegr.) B.S.G., Bryol. Eur. 3: 119. 247. 1845. *Trichostomum funale* Schwaegr., Spec. Musc. Suppl. 1(1): 150. 1811. ........ Nepal

– ***indica*** (Dix. & P. Vard.) Goffinet & Greven, J. Bryol. 22: 141. 2000. *Trigonodictyon indicum* Dix. & P. Vard. in P. Vard., Annales de Cryptogamie Exotique 1: 40. 1928. ........ E.H.; S.I.

– ***inflectens*** Mitt., J. Linn. Soc. Bot. Suppl. 1: 43. 1859. ........ Nepal

– ***khasiana*** Mitt., J. Linn. Soc. Bot. Suppl. 1: 45. 1859. ........ E.H. Endemic

– ***laevigata*** (Brid.) Brid., Bryol. Univ. 1: 183. 1826. *Campylopus laevigatus* Brid., Mant. Musc. 76. 1819. ........ W.H.; E.H.

– ***macrotheca*** Mitt., J. Linn. Soc. Bot. Suppl. 1: 44. 1859. ........ E.H. Endemic

– ***nepalensis*** Mitt., J. Linn. Soc. Bot. Suppl. 1: 46. 1859. ........ Nepal

– ***ovalis*** (Hedw.) Lindb., Act. Soc. Sc. Fenn. 10: 75. 1871. *Dicranum ovale* Hedw., Spec. Musc. 140. 1801. ........ W.H.; E.H.; S.I.; Nepal

– ***pilifera*** P. Beauv., Prodr. 58. 1805. ........ W. H.

– ***pulvinata*** (Hedw.) Sm., Engl. Bot. 24: 1728. 1867. *Fissidens pulvinatus* Timm ex Hedw., Spec. Musc. 158. 40 f. 1-3. 1801. ........ W.H.

– ***redunca*** Wils. *ex* Mitt., J. Linn. Soc. Bot. Suppl. 1: 43. 1859. ........ W.H.; E.H.

– ***somervellii*** Dix., J. Bot. 63: 221. 1925. ........ E.H.; Nepal

– ***subdoniana*** Nog., Candollea 19: 175. 1964. ........ Nepal

– ***trichophylla*** Grev., Fl. Edinensis 235. 1824. ........ E.H.

– ***unicolour*** Hook. *in* Grev., Scott. Crypt. Fl. 3: 123. 1825. ........ W.H.

***Groutiella*** Steere, Bryologist 53: 145. 1950. (Orthotrichaceae)

– ***goniorrhyncha*** (Doz. *et* Molk.) Wijk *et* Marg., Taxon 9: 50. 1960. *Schlotheimia goniorrhyncha* Doz. *et* Molk., Pl. Jungh. 3: 338. 1854. ........ E.H.; Andaman Is.

***Gymnostomiella*** Fleisch., Musci Fl. Buitenzorg 1: 309. 1904.
(Splachnaceae)

Type: *G. vernicosa* (Hook.) Fleisch.

– ***vernicosa*** (Hook.) Fleisch., Musci Fl. Buitenzorg 1: 310. 53. 1904. *Gymnostomum vernicosum* Hook., Icon. Pl. Rar. 1: 17. 1836.
........ W.H.; S.I.; C.I.; G.P.; Nepal

***Gymnostomum*** Nees *et* Hornsch., Bryol. Germ. 1: 112. 1823. (Pottiaceae)

L. Type : *G. calcareum* Nees *et* Hornsch.

– ***aeruginosum*** Sm., Fl. Brit. 3: 1163. 1804. ........W.H.; R.

– ***calcareum*** Nees *et* Hornsch., Bryol. Germ. 1: 153. 10f. 15. 1823.
........ W.H.; E.H.

***Habrodon*** Schimp., Syn. 505. 1860. (Leskeaceae)

Type: *H. notarisii* Schimp.

– ***kashmiriensis*** Vohra, Bull. Bot. Surv. India 11: 70-72. 1969.
........ W.H. Endemic

***Hageniella*** Broth., Oefv. Finsk. Vet. Soc. Foerh. 52. Afd. A(7): 4. 1910.
(Sematophyllaceae)

Type: *H. sikkimensis* Broth.

– ***assamica*** Dix., J. Bombay Nat. Hist. Soc. 39: 790. 1 f. 19. 1937.
........ E.H. Endemic

– ***isopterygioides*** Dix., J. Bombay Nat. Hist. Soc. 39: 791.1 f. 18. 1937. ........ E.H. Endemic

– ***sikkimensis*** Broth., Oefv. Finsk. Vet. Soc. Foerh. 52. Afd. A(7). 4.4. 1910. ........ E.H. Endemic

***Handeliobryum*** Broth., Sitzungsber. Ak. Wiss. Wien. Math. Nat. Kl. 133: 575. 1924. (Neckeraceae)

– ***assamicum*** Dix., J. Bombay Nat. Hist. Soc. 39: 785. Pl. 1-12. 1937.
........ E.H.

– ***himalayanum*** Broth., Sitzungsber. Ak. Wiss. Wien, Math. Nat. Kl. 133: 576. 1924. ........ E.H.

– ***setschwanicum*** Broth., Sitzungsber. Ak. Wiss. Wein, Math. Nat. Kl. 133: 575. 1924. ........ E.H.; S.I.; Nepal

***Haplocladium*** (C. Muell.) C. Muell., Nuov. Giorn. Bot. Ital. n. ser. 3: 116. 1896. (Thuidiaceae)

– ***angustifolium*** (Hamp. *et* C. Muell.) Broth., Nat. Pfl. 1(3): 1008. 1907. *Hypnum angustifolium* Hamp. *et* C. Muell., Bot. Zeit. 13: 88. 1855. ........W.H.; E.H.; Nepal

– ***himalayanum*** Bartr., Bull. Torr. Bot. Cl. 82: 27. 1955. ........ W.H.

– ***microphyllum*** (Hedw.) Broth., Nat. Pfl. 1(3): 1007. 1907. *Hypnum microphyllum* Hedw., Spec. Musc. 269. 1801. ........ E.H.; Nepal

– ***microphyllum*** (Hedw.) Broth. ssp. ***capillatum*** (Mitt.) Reim. var. *capillatum* (Mitt.) Reim., Hedwigia 76: 230. 1937. *Leskea capillata* Mitt., J. Linn. Soc. Bot. Suppl. 1. 130. 1859. ........ W.H.

– ***schimperi*** Ther., Ann. Crypt. Exot. 3: 75. 1930. ........ W.H.; E.H.

– ***stratosum*** (Mitt.) Dix., Not. R. Bot. Gard. Edinburgh 19: 295. 1938. *Leskea stratosa* Mitt., J. Linn. Soc. Bot. Suppl. 1: 131. 1859. ........ W.H.; E.H.; Nepal

***Haplohymenium*** Doz. *et* Molk., Musci Fr. ined. Archip. Indici 127. 40. 1846. (Leskeaceae)

Type: *H. sieboldii* (Doz. *et* Molk.) Doz. *et* Molk.

– ***triste*** (Ces.) Kindb., Rev. Bryol. 26: 25. 1899. *Leskea tristis* Ces. in De Not., Syll. 67. 1838. ........W.H.; Nepal

***Hedwigia*** P. Beauv., Mag. Enc. 5: 304. 1804. (Hedwigiaceae)

Type: *H. ciliata* (Hedw.) P. Beauv.

– ***ciliata*** (Hedw.) Ehrh. *ex* P. Beauv., Prodr. 15. 1805. *Anictangium ciliatum* Hedw., Spec. Musc. 40: 1801. ........ W.H.

***Hedwigidium*** B.S.G., Bryol. Eur. 3: 155. 1846. (Hedwigiaceae)

– ***integrifolium*** (P. Beauv.) Dix. *in* C. Jens., Skand. Bladmfl. 369. 1939. *Hedwigia integrifolia* P. Beauv., Prodr. 60. 1805. ........ S.I.

***Herpetineuron*** (C. Muell.) Card., Beih. Bot. Centralbl. 19(2): 127. 1905. (Leskeaceae)

Type: *H. toccoae* (Sull. *et* Lesq.) Card.

– ***toccoae*** (Sull. *et* Lesq.) Card., Beih. Bot. Centralbl. 19(2): 127. 1905. *Anomodon toccoae* Sull. *et* Lesq., Musci Bor. Am. 52. 1856. ........ W.H.; E.H.; S.I.; C.I.; Nepal

***Heterophyllium*** (Schimp.) Kindb. *in* Fleisch., Musci Fl. Buitenzorg 4: 1173. 1923. (Sematophyllaceae)

– ***confine*** (Mitt.) Fleisch., Musci Fl. Buitenzorg 4: 1177. 1923. *Stereodon confinis* Mitt., Trans. Linn. Soc. London Bot. ser. 2, 3: 1. 1891. ........ I.O.

– ***haldanianum*** (Grev.) Fleisch., Musci Fl. Buitenzorg 4: 1177. 1923. *Hypnum haldanianum* Grev., Ann. Lyc. Nat. Hist. New York 1: 275. 23. 1825. ........ W.H.

– ***renitens*** (Mitt.) Broth., Nat. Pfl. ed. 2, 11: 411. 1925. *Stereodon renitens* Mitt., Musci Ind. Or. 94. 1859. ........ E.H. Endemic

***Himantocladium*** (Mitt.) Fleisch., Musci Fl. Buitenzorg 3: 883. 1908. (Neckeraceae)

– ***cyclophyllum*** (C. Muell.) Fleisch., Musci Fl. Buitenzorg 3: 887. 1908. *Neckera cyclophylla* C. Muell., Syn. 2: 664. 1851. ........ E.H.; S.I.

– ***flagelliferum*** (Broth.) Broth., Nat. Pfl. ed. 2, 11: 190. 1925. *Neckera flagellifera* Broth. *in* Schumann *et* Lauterb., Fl. Deutsch. Schutzgeb. Sudsee 92. 1900. ........ E.H.

– ***loriforme*** (Bosch *et* Lac.) Fleisch., Musci Fl. Buitenzorg 3: 884. 1908. *Neckera loriformis* Bosch *et* Lac., Bryol. Jav. 2: 63. 183. 1863. ........ E.H.

– ***plumula*** (Nees) Fleisch., Musci Fl. Buitenzorg 3: 889. 1908. *Pilotrichum plumula* Nees *in* Brid., Bryol. Univ. 2: 759. 1827. ........ C.I.

– ***rugulosum*** (Mitt.) Fleisch., Musci Fl. Buitenzorg 3: 886. 1908. *Neckera rugulosa* Mitt., J. Linn. Soc. Bot. Suppl. 1: 121. 1859. ........ S.I.

– ***strictum*** Dix., J. Siam Soc., Nat. Hist. Suppl. 9(1): 30. 1932. ........ S.I.

***Holomitrium*** Brid., Bryol. Univ. 1: 226. 1826. (Dicranaceae)

- ***densifolium*** (Wils.) Wijk & Marg., Taxon 11: 221. 1962. *Symblepharis densifolia* Wils., Kew J. Bot. 9: 292. 1857. ........ E.H.; S.I.

***Homalia*** (Brid.) B.S.G., Bryol. Eur. 5: 53. 1850. (Neckeraceae)

Type : *H. trichomanoides* (Hedw.) B.S.G.

- ***obtusata*** Mitt., J. Linn. Soc. Bot. 8: 38. 1864. ........ W.H.
- ***pygmaea*** Broth., Nat. Pfl. 1(3). 849. 1906. ....... S.I.

***Homaliadelphus*** Dix. *et* P. Vard., Rev. Bryol. n. ser. 4: 142. 1932. (Neckeraceae)

Type: *H. targionianus* (Mitt.) Dix. *et* P. Vard.

- ***targionianus*** (Mitt.) Dix. *et* P. Vard., Rev. Bryol. n. ser. 4: 142. 1932. *Neckera targioniana* Mitt., J. Linn. Soc. Bot. Suppl. 1: 117. 1859. ........ S.I.

***Homaliodendron*** Fleisch., Hedwigia 45: 74. (I) 1906. (Neckeraceae)

- ***crassinervium*** Ther., Recueil Publ. Soc. Havraise, Etud. Div. 1919: 39. 2f. 2. 1919.
- ***dentatum*** (Griff.) Fleisch., Hedwigia 45:75. (I) 1906. *Neckera dentata* Griff., Notul. Pl. As. 463. 1849. *et* Icon. Pl. As. 88*f.* 2. 1849. ........ E.H.
- ***exiguum*** (Bosch *et* Lac.) Fleisch., Musci Fl. Buitenzorg 3: 897. 156. 1908. *Homalia exigua* Bosch *et* Lac., Bryol. Jav. 2: 55. 175. 1862. ........ W.H.; E.H.; S.I.
- ***flabellatum*** (Sm.) Fleisch., Hedwigia 45: 74. 1906. *Hookeria flabellata* Sm., Trans. Linn. Soc. 9: 280. 23 *f.* 2. 1808. ........ E.H.
- ***javanicum*** (C. Muell.) Fleisch., Hedwigia 45: 74. (I) 1906. *Neckera javanica* C. Muell., Syn. 2: 41. 1850. ........ C.I.
- ***ligulaefolium*** (Mitt.) Fleisch., Hedwigia 45: 77. 1906. *Neckera ligulaefolia* Mitt., J. Linn. Soc. Bot. Suppl. 1: 119. 1859. ........ E.H.
- ***microdendron*** (Mont.) Fleisch., Hedwigia 45: 78. 1906. *Hookeria microdendron* Mont., Ann. Sc. Nat. Bot. ser. 2, 19: 240. 1843. ........ W.H.; E.H.; S.I.

– ***montagneanum*** (C. Muell.) Fleisch., Hedwigia 45: 74. 1906. *Neckera montagneana* C. Muell., Bot. Zeit. 14: 436. 1856. ........ E.H.; S.I.; Nepal

– ***obtusatum*** (Mitt.) Gang., Mosses E. India 5: 1416. 1976. *Homalia obtusata* Mitt., J. Linn. Soc. Bot. 8: 38. 1864. ........ W.H.; E.H.

– ***paquei*** (Ren. *et* Card.) Broth., Nat. Pfl. 1(3): 851. 1906. *Homalia paquei* Ren. *et* Card., Bull. Soc. R. Bot. Belg. 41(1): 81. 1905. ........ E.H.

– ***rectifolium*** (Mitt.) Fleisch., Hedwigia 45: 75. 1906. *Neckera rectifolia* Mitt., J. Linn. Soc. Bot. Suppl. 1: 119. 1859. ........ E.H.; Nepal

– ***scalpellifolium*** (Mitt.) Fleisch., Hedwigia 45: 75. 1906. *Neckera scalpellifolia* Mitt., J. Linn. Soc. Bot. Suppl. 1: 119. 1859. ........ W.H.; E.H.; Nepal

– ***sphaerocarpum*** Nog., Musci-Fl. East. Himalayas 576. 1966. ........ E.H.

– ***stracheyanum*** (Mitt.) Fleisch., Hedwigia 45: 78. 1906. *Neckera stracheyana* Mitt., J. Linn. Soc. Bot. Suppl. 1: 118. 1859. ........ W.H.

***Homalothecium*** B.S.G., Bryol. Eur. 5: 91. 1851. (Brachytheciaceae)

Type: *H. sericeum* (Hedw.) B.S.G.

– ***euchloron*** (Bruch ex C. Muell.) R.S. Chopra, Taxon. Indian Mosses: 460. 1975. *Hypnum euchloron* Bruch ex C. Muell., Syn. Musc. 2: 464. 1851. ........ W.H.

– ***incompletum*** (Griff.) Jaeg. , Ber. S. Gall. Naturw. Ges. 1877-78: 482. 1880. *Leskea incompleta* Griff., Notul. Pl. As. 2: 478. 1849 & Icon. Pl. A. 1849. ........ E.H. Endemic

– ***integerrimum*** Dix., Ann. Bryol. 3: 68. 1930. ........ W.H. Endemic

– ***neckeroides*** (Griff.) Par., Ind. Bryol. 568. 1896. *Pterogonium neckeroides* Griff., Calcutta J. Nat. Hist. 3: 64. 1843. ........ E.H. Endemic

– ***nilgheriense*** (Mont.) Robins., Bryologist 65: 99. 1962. *Isothecium neilgheriense* Mont., Ann. Sc. Nat. Bot. ser. 2, 17: 246. 1842. ........ W.H.; E.H.; S.I.; Nepal

– ***sericeum*** (Hedw.) B.S.G., Bryol. Eur. 5: 93. 456. 1851. *Leskea sericea* Hedw., Spec. Musc. 228. 1801. ........ W.H.; Nepal

***Homomallium*** (Schimp.) Loesk., Hedwigia 46: 314. 1907. (Hypnaceae)

– ***incurvatum*** (Brid.) Loesk., Hedwigia 46: 314. 1907. *Hypnum incurvatum* Schrad. *ex* Brid., Musc. Rec. 2(2): 119. 1801. ........ W.H.

– ***simalaense*** (Mitt.) Broth., Nat. Pfl. 1(3): 1027. 1908. *Stereodon simalensis* Mitt., J. Linn. Soc. Bot. Suppl. 1: 95. 1859. ........ W.H.

***Hookeria*** J.E. Sm., Trans. Linn. Soc. London 9: 275. 1808. (Hookeriaceae)

L. Type : *H. lucens* (Hedw.) Sm.

– ***acutifolia*** Hook. *et* Grev., Edinburgh J. Sc. 2: 225. 5. 1825. ........ W.H.; E.H.; S.I.; Nepal

***Hookeriopsis*** (Besch.) Jaeg., Ber. S. Gall. Naturw. Ges. 1875-76: 358. 1877. (Hookeriaceae)

– ***secunda*** (Griff.) Broth., Nat. Pfl. 1(3): 942. 1907. *Hookeria secunda* Griff., Calcutta J. Nat. Hist. 3: 280. 1843. ........ E.H. Endemic

– ***utacamundiana*** (Mont.) Broth., Nat. Pfl. 1(3): 942. 1907. *Hookeria utacamundiana* Mont., Ann. Sc. Nat. Bot. ser. 2, 17: 247. 1842. ........ E.H.; S.I.

***Horikawaea*** Nog., J. Sc. Hiroshima Univ. B(2) Bot. 3: 47. 6. 1937. (Pterobryaceae)

Type: *H. nitida* Nog.

– ***nitida*** Nog., J. Sc. Hiroshima Univ. B(2) Bot. 3: 47. 6. 1937. ........ E.H.

***Hydrocryphaea*** Dix., J. Bot. 69: 3. 1931. (Cryphaeaceae)

Type: *H. wardii* Dix.

– ***wardii*** Dix., J. Bot. 69: 4. 595 f. 3. 1931. ........ E.H.

***Hydrogonium*** (C. Muell.) Jaeg., Ber. S. Gall. Naturw. Ges. 1877-78: 405. 1880. (Pottiaceae)

- ***amplexifolium*** (Mitt.) Chen, Hedwigia 80: 240.46 f. 3-5. 1941. *Tortula amplexifolia* Mitt., J. Linn. Soc. Bot. Suppl. 1: 29. 1859. ........ W.H.; E.H.

- ***arcuatum*** (Griff.) Wijk *et* Marg., Taxon 7: 289. 1958. *Barbula arcuata* Griff., Calcutta J. Nat. Hist. 2: 491. 1842. ........ W.H.; E.H.; S.I.; C.I..; G.P.; Nepal

- ***consanguineum*** (Thwait. *et* Mitt.) Hilp., Beih. Bot. Centralbl. 50(2): 626. 1933. *Tortula consanguinea* Thwait. *et* Mitt., J. Linn. Soc. 13: 300. 1873. ........ W.H.; E.H.; S.I.; C.I.; G.P.; Nepal

- ***decolyi*** Gang., Nov. Hedwigia 12: 426. 1966. ........ E.H. Endemic

- ***dicranelloides*** Gang., Nov. Hedwigia 12: 427. 1966. ........ W.H.; E.H.

- ***ehrenbergii*** (Lor.) Jaeg., Ber. S. Gall. Naturw. Ges. 1877-78: 405. 1880. ........ W.H.; Nepal

- ***gangeticum*** (C. Muell.) Chen, Hedwigia 80: 237. 1941. *Barbula gangetica* C. Muell., Linnaea 37: 177. 1872. ........ W.H.; G.P.

- ***gracilentum*** (Mitt.) Chen, Hedwigia 80: 237. 44 f.1, 2. 1941. *Barbula gracilenta* Mitt., Musci Ind. Or. 35. 1859. ........ W.H.; E.H.; C.I. Endemic

- ***javanicum*** (Doz. *et* Molk.) Hilp., Beih. Bot. Centralbl. 50(2): 632. 6c. 1933. *Barbula javanica* Doz. *et* Molk., Ann. Sc. Nat. Bot. ser. 3, 2: 300. 1844. ........ W.H.; E.H.; P.R.; G.P.; Nepal

- ***mussoorianum*** Vohra, J. Bombay Nat. Hist. Soc. 63(2): 464. 1967. ........ W.H.

- ***pseudo-ehrenbergii*** (Fleisch.) Chen, Hedwigia 80: 242. 47 f. 2-5. 1941. *Barbula pseudo-ehrenbergii* Fleisch., Musci Fl. Buitenzorg 1: 356. 1904. ........ W.H.; S.I.; Nepal

- ***subpellucidum*** (Mitt.) Hilp., Beih. Bot. Centralbl. 50(2) 622. 6b, 7b. 1933. *Barbula subpellucida* Mitt., Musci Ind. Or. 35. 1859. ........ W.H.; E.H.

***Hygroamblystegium*** Loesk., Moosfl. Harz. 298. 1903. (Amblystegiaceae)

Type: *H. irriguum* (Hook. *et* Wils.) Loesk.

- ***gangulianum*** Vohra, Bull. Bot. Surv. India 22: 115. 1980. ........ W.H. Endemic

– ***obtusulum*** (Mitt.) Broth., Nat. Pfl. 1(3): 1028. 1908. *Hypnum obtusulum* Mitt., J. Linn. Soc. Bot. Suppl. 1: 82. 1859. ........ W.H.; Nepal

– ***tenax*** (Hedw.) Jenn., Man. Moss. W. Pennsylv. 277. 39. 1913. *Hypnum tenax* Hedw., Spec. Musc. 277. 1801. ........ W.H.

***Hygrohypnum*** Lindb., Act. Soc. Sc. Fenn. 10: 277. 1872. (Amblystegiaceae)

L. Type: *H. palustre* Loesk.

– ***choprae*** Vohra, Bull. Bot. Surv. India 22: 117. 1980. ........ E.H. Endemic

– ***dilatatum*** (Wils.) Loesk., Moosfl. Harz. 320. 1903. *Hypnum dilatatum* Wills. in Schimp., Syn. ed. 2: 776. 1876. ........ W.H.

– ***luridum*** (Hedw.) Jenn., Man. Moss. W. Pennsylv. 287. 1913. *Hypnum luridum* Hedw., Spec. Musc. 291. 1801. ........ W.H.

– ***nairii*** Vohra, Bull. Bot. Surv. India 22: 117. 1980. ........ W.H. Endemic

***Hylocomium*** B.S.G., Bryol. Eur. 5: 169. 1852. (Hylocomiaceae)

L. Type : *H. splendens* (Hedw.) B.S.G.

– ***himalayanum*** (Mitt.) Jaeg., Ber. S. Gall. Naturw. Ges. 1877-78: 348. 1880. *Stereodon himalayanus* Mitt., Musci Ind. Or. 113. 1859. ........ W.H.; E.H.; Nepal

– ***indicum*** Dix., Not. R. Bot. Gard. Edinburgh 19: 299. 13. 1938. ........ W.H.

– ***splendens*** (Hedw.) B.S.G., Bryol. Eur. 5: 173. 487. 1852. *Hypnum splendens* Hedw., Spec. Musc. 262. 1801. ........ Nepal

***Hymenostomum*** R. Brown, Trans. Linn. Soc. London 12(2): 572. 1819. (Pottiaceae)

Type: *H. microstomum* (Hedw.) R. Brown

– ***edentulum*** (Mitt.) Besch., Bull. Soc. Bot. France 34: 95. 1887. *Weisia edentula*.Mitt., Musci Ind. Or. 27. 1859. ........ S.I.; C.I.

– ***microstomum*** (Hedw.) R. Brown in Nees *et* Hornsch., Bryol. Germ. 1: 196. 12 f. 4. 1823. *Gymnostomum microstomum* Hedw., Spec. Musc. 33. 1801. ........ W.H.

– ***rostellatum*** (Brid.) Schimp., Syn. ed. 2: 33. 1876. ........ W.H.

– ***tortile*** (Schwaegr.) B.S.G., Bryol. Eur. 1: 56. 18. 1846. *Gymnostomum tortile* Schwaegr. *in* Schkuhr, Deutschl. Krypt. Gew. 2(1): 28. 1810. ........ W.H.

***Hymenostyliella*** Bartr., Philipp. J. Sc. 68: 108. 1939. (Pottiaceae)

Type: *H. involuta* (Card. *et* Ther.) Bartr.

– ***llanosii*** (Broth.) Robins., Phytologia 21: 2. 1971. ........ W.H.; C.I.

***Hymenostylium*** Brid., Bryol. Univ. 2: 81. Suppl. 3. 1827. (Pottiaceae)

Type: *H. xanthocarpum* (Hook.) Brid.

– ***annotinum*** Mitt. *ex* Dix., J. Bot. 48: 308. 11. 1910. ........ I.O.

– ***dicranelloides*** Broth. *ex* Dix., Bryologist 30: 109. 1928. ........ W.H.

– ***filiforme*** Dix., Ann. Bryol. 3: 54. 1930. ........ W.H.

– ***grandirete*** Dix., 150th Anniv. R. Bot. Gard. Calcutta 178. 1942. ........ W.H.

– ***recurvirostrum*** (Hedw.) Dix., Rev. Bryol. Lich. 6: 96. 1934. *Gymnostomum recurvirostrum* Hedw., Spec. Musc. 33. 1801. ........ W.H.; E.H.; S.I.; P.P.; Nepal

var. ***aurantiacum*** (Mitt.) Gang., Mosses E. India 648. 1972. *H. aurantiacum* Mitt., J. Linn. Soc. Bot. Suppl. 1: 32. 1859. ........ W.H.; E.H.; S.I.; Nepal

– ***shepheardae*** Card. *et* Dix., J. Bot. 48: 307.10. 1910. ........ W.H.

– ***validinerve*** Dix. *et* P. Vard., Ann. Crypt. Exot. 3: 174. 2f. 3. 1930. ........ S.I.

***Hyophila*** Brid., Bryol. Univ. 1 Suppl. : 760. 1827. (Pottiaceae)

L. Type : *H. javanica* (Nees *et* Blum.) Brid.

– ***comosa*** Dix. *et* P. Vard., Arch. Bot. 1(8-9): 166.5 f. 2. 1927. ........ S.I.; C.I. Endemic

– ***involuta*** (Hook.) Jaeg., Ber. S. Gall. Naturw. Ges. 1871-72: 354. 1873. *Gymnostomum involutum* Hook., Musc. Exot. 2: 154. 1819. ........ W.H.; E.H.; R.; S.I.; C.I.; G.P.; Nicobar Is.; Nepal

– ***kurziana*** Gang., Nov. Hedwigia 12: 422. 1966. ........ E.H. Endemic

– ***perannulata*** Ren. *et* Card., Bull. Soc. R. Bot. Belg. 34(2): 60. 1896. ........ E.H. Endemic

– ***rosea*** Williams, Bull. New York Bot. Gard. 8: 341. 1914. ........ W.H.; S.I.; C.I.; G.P.

– ***spathulata*** (Harv.) Jaeg., Ber. S. Gall. Naturw. Ges. 1871-72: 353. 1873. *Gymnostomum spathulatum* Harv. *in* Hook., Icon. Pl. Rar. 1: 17. 1836. ........ W.H.; E.H.; Nepal

***Hyophilopsis*** Card. *et* Dix., J. Bot. 49: 137. 1911. (Pottiaceae)

Type: *H. entosthodontacea* Card. *et* Dix.

– ***entosthodontacea*** Card. *et* Dix., J. Bot. 49: 137. 513 f. 1. 1911. ........ S.I.

***Hypnum*** Hedw., Spec. Musc. 236. 1801. (Hypnaceae)

L. Type : *H. cupressiforme* Hedw.

– ***aduncoides*** (Brid.) C. Muell. Syn. 2: 295. 1851. *H. cupressiforme* Hedw. var. *aduncoides* Brid., Spec. Musc. 2: 219. 1812. ........ E.H.

– ***cupressiforme*** L. *ex* Hedw., Spec. Musc. 291. 1801. ssp. ***imponens*** (Hedw.) Boul., Muscin. France 1: 34. 1884. ........ W.H.; E.H.; S.I.

– ***flaccens*** Besch., Ann. Sc. Nat. Bot. ser. 7. 15: 92. 1892. ........ E.H.; Nepal

– ***macrogynum*** Besch., Ann. Sc. Nat. Bot. ser. 7.15: 91. 1892. ........ E.H.; Nepal

– ***plumaeforme*** Wils., London J. Bot. 7: 277. 1848. ........ Nepal

– ***setschwanicum*** (Broth.) Ando, J. Sc. Hiroshima Univ. ser. B(2). 16(1): 24. 1976. *Stereodon setschwanicus* Broth. *in* Handel-Mazzetti, Symb. Sin. 4: 123. 1929. ........ Nepal

– ***sikkimense*** Ando, J. Jap. Bot. 43: 171. 1968. ........ E.H.; Nepal

– ***subimponens*** Lesq., Trans. Am. Phil. Soc. n. ser. 13: 14. 1865. ssp. ***ulophyllum*** (C. Muell.) Ando, Bot. Mag. Tokyo 79: 766. 1966. *Cupressina ulophylla* C. Muell., Nuov. Giorn. Bot. Ital. n. ser. 5: 196. 1898. ........ E.H.; Nepal

– ***submolluscum*** Besch., Ann. Sc. Nat. Bot. ser. 7. 15: 93. 1892. ........ E.H.

– ***vaucheri*** fo. ***vaucheri*** Lesq., Mem. Soc. Sc. Nat. Neuchatel 3(3): 48. 1846. ........ W.H.; Nepal

***Hypopterygium*** Brid., Bryol. Univ. 2: 709. 1827. (Hypopterygiaceae)

– ***aristatum*** Bosch & Lac., Bryol. Jav. 2: 12, 141. 1861. ........ Himalayas

– ***flavolimbatum*** C. Muell., Syn. 2: 10. 1850. ........ W.H.; E.H.; Nepal

– ***tenellum*** C. Muell., Bot. Zeit. 12: 557. 1854. ........ S.I.

***Isopterygiopsis*** Iwats., J. Hattori Bot. Lab. 33: 379. 1970. (Hypnaceae)

– ***muelleriana*** (Schimp.) Iwats., J. Hattori Bot. Lab. 33: 379. 1970. *Plagiothecium muellerianum* Schimp., Syn. 584. 1860. ........ W.H.; E.H.; Nepal

***Isopterygium*** Mitt., J. Linn. Soc. Bot. 12: 21. 497. 1969. (Hypnaceae)

L. Type : *I. planissimum* Mitt.

– ***albescens*** (Hook.) Jaeg., Ber. S.Gall. Naturw. Ges. 1876-77: 433. 1878. *Hypnum albescens* Hook. *in* Schwaegr., Spec. Musc. Suppl. 3(1): 226 b. 1828. ........ W.H.; E.H.; Nepal

var. ***smallii*** (Sull. & Lesq.) Iwats., J. Hattori Bot. Lab. 29: 62. 1966. *Hypnum smallii* Sull.& Lesq., Proc. Am..Ac. Arts. Sc. 4: 281. 1859. ........ W.H.; E.H.; S.I.; Nepal

– ***andamanicum*** Gang., Mosses E. India 8: 1953. 1980. ........ Andaman Is. Endemic

– ***assamicum*** (Mitt.) Jaeg., Ber. S. Gall. Naturw. Ges. 1876-77: 432. 1878. *Stereodon assamicus* Mitt., Musci Ind. Or. 104. 1859. ........ E.H.

– ***bancanum*** (Lac.) Jaeg., Ber. S. Gall. Naturw. Ges. 1876-77: 442. 1878. *Hypnum bancanum* Lac., Bryol. Jav. 2: 188, 286. 1868. ........ I.O.

– ***distichaceum*** (Mitt.) Jaeg., Ber. S. Gall. Naturw. Ges. 1876-77: 439. 1878. *Stereodon distichaceus* Mitt., Musci Ind. Or. 105. 1859. ........ W.H.; E.H.; Nepal

– ***elegans*** (Brid.) Lindb., Not. Saellsk. F. Fl. Fenn. Foerh. 13: 416. 1874. *Isothecium elegans* Brid., Bryol. Univ. 2: 366. 1827. ........ W.H.; E.H.; Nepal

– ***lignicola*** (Mitt.) Jaeg., Ber. S. Gall. Naturw. Ges. 1876-77: 432. 1878. *Hypnum lignicola* Mitt., Kew J. Bot. 8: 355. 1856. ........ E.H.; S.I.

– ***longitheca*** (Mitt.) Jeag., Ber. S. Gall. Naturw. Ges. 1876-77. 432. 1878. *Stereodon longitheca* Mitt., Musci Ind. Or. 104. 1859. ........ E.H. Endemic

– ***micans*** (Sw.) Kindb., Enum. Bryin. Exot. 21. 1888. *Hypnum micans* Sw., Adnot Bot. 175. 1829. ........ E.H.

– ***microplumosum*** (C. Muell.) Broth., Nat. Pfl. 1(3): 1083. 1908. *Taxicaulis microplumosus* C. Muell., Hedwigia 40: 68. 1901. ........ Andaman Is.

– ***minutirameum*** (C. Muell.) Jaeg., Ber. S. Gall. Naturw. Ges. 1876-77: 434. 1878. *Hypnum minutirameum* C.Muell., Syn. 2: 689. 1851. ........ W.H.; E.H.; S.I.

– ***muellerianum*** (Schimp.) Jaeg., Ber. S. Gall. Naturw. Ges. 1876-77: 441. 1878. *Plagiothecium muellerianum* Schimp., Syn. 584. 1860. ........ W.H.

– ***pallidulum*** (Mitt.) Jaeg., Ber. S. Gall. Naturw. Ges. 1876-77: 432. 1878. *Stereodon pallidulus* Mitt., Musci Ind. Or. 105. 1859. ........ W.H.; E.H.

– ***pohliaecarpum*** (Sull. & Lesq.) Jaeg., Ber. S. Gall. Naturw. Ges. 1876-77 : 442. 1878. *Hypnum pohliaecarpum* Sull. *et* Lesq., Proc. Am. Ac. Arts. Sc. 4: 280. 1859. ........ E.H.; S.I.

– ***pulchellum*** (Hedw.) Jaeg., Ber. S. Gall. Naturw. Ges. 1876-77: 441. 1878. *Leskea pulchella* Hedw., Spec. Musc. 220. 55f. 7-12. 1801. ........ W.H.

– ***seligeri*** (Brid.) Dix. *in* C. Jens., Skand. Bladm. Fl. 489. 1939. *Leskea seligeri* Brid., Musc. Rec. 2(2): 47. 1801. ........ E.H.

– ***serrulatum*** Fleisch., Musci Fl. Buitenzorg 4: 1433. 1923. ........ E.H. Endemic

– ***subalbescens*** Broth., Hedwigia 38: 231. 1899. ........ E.H.; S.I.

– ***textorii*** (Lac.) Mitt., Tran. Linn. Soc. London. Bot. ser. 2. 3: 176. 1891. *Rhynchostegium textorii* Lac., Ann. Mus. Bot. Lugd. Bot. 2: 299. 1867. ........ E.H.; S.I.

– ***undulatum*** Dix. *et* P. Vard., Ann. Crypt. 1: 461. 1928. ........ S.I.

***Isothecium*** Brid., Bryol. Univ. 2: 355. 10. 1827. (Lembophyllaceae)

L. Type : *I. viviparum* Lindb.

– ***comes*** (Griff.) Nog., Bryologist 73: 134. 1970. *Neckera comes* Griff., Calcutta J. Nat. Hist. 3: 71. 1843. ........ E.H.

– ***semitortum*** Dix., Not. R. Bot. Gard. Edinburgh 19: 294. 10. 1938. ........ Nepal

***Jaegerina*** C. Muell., Linnaea 40 : 273. 1876. (Pterobryaceae)

Type: *J. stolonifera* (C. Muell.) C.Muell.

– ***stolonifera*** (C.Muell.) C. Muell., Linnaea 40: 273. 1876. *Pilotrichum stoloniferum* C. Muell., Bot. Zeit. 20: 382. 1862. ........ S.I.

***Juratzkaea*** Lor., Bot. Zeit. Regensburg 24: 188. 1866. (Plagiotheciaceae)

Type : *J. seminervis* (Schwaegr.) Lor.

– ***indica*** Broth. *et* P. Vard., Rev. Bryol. 52: 40. 1925. ........ S.I.

***Kashyapia*** R.S. Chopra, Taxon. Indian Mosses: 297. 1975. (Hedwigiaceae)

– ***ambigua*** (Hook.) R.S. Chopra, Taxon. Indian Mosses: 298. 1975. ........ W.H.; E.H.; Nepal

***Leiodontium*** Broth., Symb. Sin. 4: 127. 1929. (Hypnaceae)

– ***complanatum*** Dix., J. Bombay Nat. Hist. Soc. 39(4): 794. 1937. ........ E.H. Endemic

– ***gracile*** Broth., Symb. Sin. 4: 127. 1929. ........ W.H.; Nepal

***Lepidopilidium*** (C.Muell.) Brottl., Nat. Pfl. 1(3): 942. 1907. (Hookeriaceae)

– ***furcatum*** (Thwait. et Mitt.) Broth., Nat. Pfl. 1(3): 944. 1907. *Lepidopilum furcatum* Thwait. et Mitt., J. Linn. Soc. Bot. 13: 310. 1873. ........ S.I.

***Leptobryum*** (B.S.G.) Wils., Bryol. Brit. 219. 1855. (Bryaceae)

– ***pyriforme*** (Hedw.) Wils., Bryol. Brit. 219. 1855. *Webera pyriformis* Hedw., Spec. Musc. 169. 1801. ........ W.H.; E.H.

***Leptodictyum*** (Schunp.) Warnst., Kryptog. Fl. Brandenburg 2: 840, 867. 1906. (Amblystegiaceae)

– ***riparium*** (Hedw.) Warnst., Kryptog. Fl. Brandenburg 2: 878. 1906. *Hypnum riparium* Hedw., Spec. Musc. 241. 1801. ........ W.H.

***Leptodontium*** (C. Muell.) Hamp., Oefv. K. Vet. Ak. Foerh. 21: 227. 1864. (Pottiaceae)

Type : *L. squarrosum* (Hook.) Hamp. in Lindb.

– ***flexifolium*** (Dicks.) Hamp. *in* Lindb., *Bryum flexifolium* Dicks. ex With., Syst. Arr. Brit. Pl. ed. 4, 3: 799. 1801. ........ W.H.; E.H.; Nepal

– ***handelii*** Ther., Ann. Crypt. Exot. 5: 171. 1932. ........ E.H.

– ***viticulosoides*** (P. Beauv.) Wijk *et* Marg., Taxon 9: 51. 1960. *Neckera viticulosoides* P. Beauv., Prodr. 75. 1805. ........ E.H.; Nepal

***Leptohymenium*** Schwaegr., Spec. Musc. Suppl. 3(1): 246 c. 1828. (Hylocomiaceae)

Type: *L. tenue* (Hook.) Schwaegr.

– ***tenue*** (Hook.) Schwaegr., Spec. Musc. Suppl. 3(1): 246 c. 1828. *Neckera tenuis* Hook., Trans. Linn. Soc. London 9: 315. 27(3). 1808. ........ W.H.; E.H.; Nepal

***Leptopterigynandrum*** C. Muell., Hedwigia 36: 114. 1897. (Leskeaceae)

Type: *L. austro-alpinum* C. Muell.

– ***autoicum*** Dix. *ex* Gang. *et* Vohra, Mosses E. India 7: 1582. 1978. ........ I.O.

– ***brevirete*** Dix. *in* Blatter *et* Fernandez, J. Indian Bot. Soc. 10(2): 151. 1931. ........ W.H.

– ***decolor*** (Mitt.) Fleisch., Musci Fl. Buitenzorg 4: 1496. 1923. *Stereodon decolor* Mitt.. J.Linn. Soc. Bot. Suppl. 1: 92. 1859. ........ E.H. Endemic

– ***subintegrum*** (Mitt.) Broth., Nat. Pfl. ed. 2. 11: 309. 1925. *Heterocladium subintegrum* Mitt., J.Linn. Soc. Bot. Suppl. 1: 135. 1859. ........ W.H.; Nepal

***Lescuraea*** B.S.G., Bryol. Eur. 5 : 101. 1851. (Leskeaceae)

Type: *L. striata* (Schwaegr.) B.S.G.

– ***darjeelingensis*** Vohra, Bull. Bot. Surv. India 20: 150-152. 1978. ........ E.H.

– ***incurvata*** (Hedw.) Lawt., Bull. Torr. Bot. Cl. 84: 290. 1957. *Leskea incurvata* Hedw., Spcc. Musc. 216. 1801. ........ W.H.; Nepal

– ***laevifolia*** (Mitt.) R.S. Chopra, Taxon. Indian Mosses: 421. 1975. *Leskea laevifolia* Mitt., J. Proc. Linn. Soc. Bot. Suppl. 1: 130. 1859. ........ W.H.

– ***mutabilis*** (Brid.) Lindb. *ex* Hag., Norsh. Vid. Selsk. Skrift. 1908(9): 53. 1909. *Hypnum mutabile* Brid., Musc. Rec. 2(2): 170. 6 f.1.1801. ........ W.H.

– ***ramuligera*** (Mitt.) R.S. Chopra, Taxon. Indian Mosses: 421. 1975. *Leskea ramuligera* Mitt., Musci Ind. Or. 132. 1859. ........ W.H.; Nepal

– ***saxicola*** (B.S.G.) Mol. *in* Lor., Moss Stud. 144.147.149. 1864. *L. striata* (Schwaegr.) B.S.G., var. *saxicola* B.S.G., Bryol. Eur. 5: 103-459. 1851. ........ W.H.

***Leskea*** Hedw., Spec. Musc. 211. 1801. (Leskeaceae)

L. Type : *L. polycarpa* Hedw.

– ***consanguinea*** (Mont.) Mitt., J. Linn. Soc. Bot. Suppl. 1: 131. 1859. *Pterogonium consanguineum* Mont., Ann. Sc. Nat. Bot. ser. 2, 17: 249. 1842. ........ S.I.

– ***hyaloapiculata*** Dix., Not. R. Bot. Gard. Edinburgh 19: 294. 1938. ........ W.H.

– ***perstricta*** Dix., Rec. Bot. Surv. India 6(3): 68. 1914. ........ E.H. Endemic

***Leskeella*** (Limpr.) Loesk., Mosfl. Harz. 255. 1903. (Leskeaceae)

Type: *L. nervosa* (Brid.) Loesk.

– ***incrassata*** (Broth.) Broth., Nat. Pfl. 1(3): 996. 1907. *Leskea incrassata* Lindb. *ex* Broth., Ac. Soc. Sc. Fenn. 19(12): 96. 1892. ........ W.H.

– ***nervosa*** (Brid.) Loesk., Moosfl. Harz. 255. 1903. *Pterigynandrum nervosum* Brid., Spec. Musc. 1: 132. 1806. ........ W.H.; P.

***Leucobryum*** Hamp., Linnaea 13: 42. 1839. (Leucobryaceae)

Type: *L. vulgare* Hamp.

– ***aduncum*** Doz. *et* Molk., Pl. Jungh. 3: 319. 1854. ........ W.H.; E.H.; S.I.; Nepal

– ***bowringii*** Mitt., J. Linn. Soc. Bot. Suppl. 1: 26. 1859. ........ E.H.; S.I.; Nicobar Is.

– ***cucullifolium*** Card. *in* Ren. *et* Card., Bull. Soc. R. Bot. Belg. 41: 30. 1905. ........ W.H.; E.H.; Nepal

– ***humillimum*** Card., Mem. Soc. Natn. Sc. Math. Cherbourg 32: 15. 29. 1901. ........ S.I.; P.

– ***imbricatum*** Broth., Rec. Bot. Surv. India 1(12): 314. 1898. ........ S.I.

– ***javense*** (Brid.) Mitt., Musci Ind. Or. 25. 1859. *Sphagnum javense* Brid., Bot. Zeit. Regensburg 1: 200. 1802. ........ W.H.; E.H.; Andaman Is.; Nepal

– ***juniperoideum*** (Brid.) C. Muell., Linnaea 18: 689. 1845. *Dicranum juniperoideum* Brid., Bryol. Univ. 1: 409. 1826. ........ E.H.; S.I.

– ***mittenii*** Besch., J. de Bot. 12: 287. 1898. ........ E.H.

– ***nilghiriense*** C.Muell., Bot. Zeit. 32: 556. 1854. ........ W.H.; E.H.; S.I.; Nepal

– ***sanctum*** (Brid.) Hamp., Linnaea 13: 42. 1839. *Dicranum glaucum* Hedw. var. *sanctum* Brid., Bryol. Univ. 1: 811. 1826. ........ E.H.; Nepal

– ***scalare*** C.Muell. *ex* Fleisch., Musc. Buitenzorg 1: 143. 1904. ........ E.H.; S.I.; Andaman Is.; Nepal

– ***textorii*** Besch., J. de Bot. 12: 288. 1898. ........ S.I.

***Leucodon*** Schwaegr. *in* Hedw., Spec. Musc. Suppl. 12: 1. 1816. (Leucodontaceae)

L. Type : *L. sciuroides* (Hedw.) Schwaegr.

– ***sciuroides*** (Hedw.) Schwaegr., Spec. Musc. Suppl. 1(2): 1. 1816. *Fissidens sciuroides* Hedw., Spec. Musc. 161. 1801. ........ W.H.

– ***secundus*** (Harv. *in* Hook.) Mitt., J. Linn. Soc. Bot. Suppl. 1: 124. 1859. *Sclerodontium secundum* Harv. in Hook., Icon. Pl. Rar. 1: 21. 1836. ........ W.H.; E.H.; S.I.; Nepal

***Leucoloma*** Brid., Bryol. Univ. 2: 218. 1827. (Dicranaceae)

Type: *L. bifidum* (Brid.) Brid.

– ***amoene-virens*** Mitt., Musci Ind. Or. 13. 1859. ........ E.H.; S.I.

– ***brevifolium*** Dix. *et* P. Vard., Ann. Crypt. Exot. 3(4): 170. f. 1. 1930. ........ S.I.

– ***malabarense*** Besch. *ex* Ren. *et* Card., Bull. Soc. R. Bot. Belg. 41(1): 16. 1905. ........ S.I.

– ***molle*** (C. Muell.) Mitt., J. Linn. Soc. Bot. Suppl. 1: 13. 1859. *Dicranum molle* C. Muell., Syn. 1: 354. 1848. ........ S.I.

– ***nitens*** (Thwait. *et* Mitt.) Jaeg., Ber. S. Gall. Naturw. Ges. 1870-71: 412. 1872. *Poecilophyllum nitens* Thwait. *et* Mitt., J. Linn. Soc. Bot. 13: 297. 1873. ........ S.I.

– ***renauldii*** Broth., Rec. Bot. Surv. India 1(12): 313. 1898. ........ S.I.

– ***strictifolium*** Dix., J. Indian Bot. 2: 176. 1921. ........ S.I.

– ***taylorii*** (Schwaegr.) Mitt., Musci Ind. Or. 13. 1859. *Syrrhopodon taylorii* Schwaegr., Spec. Musc. Suppl. 2(2): 115. 1824. ........ Nepal

– ***tennerum*** Mitt.,.Kew J. Bot. 8: 353. 1856. ........ S.I.

– ***walkeri*** Broth., Rec. Bot. Surv. India 1: 313. 1899. ........ S.I.

***Leucomium*** Mitt., J. Linn. Soc. Bot. 10: 181. 1868. (Hypnaceae)

Type: *L. debile* (Sull.) Mitt.

– ***aneurodictyon*** (C.Muell.) Jaeg., Ber. S. Gall. Naturw. Ges. 1877-78: 275. 1880. *Hypnum aneurodictyon* C. Muell., Syn. 2: 681. 1851. ........ S.I.

– ***decolyi*** Broth. *ex* Gang., Mosses E. India 6: 1528. 1977. ........ E.H. Endemic

***Leucophanes*** Brid., Bryol. Univ. 1: 763. 1826. (Calymperaceae)

L. Type : *L. octoblepharioides* Brid.

– ***albescens*** C. Muell., Bot. Zeit. 22: 347. 1864. ........ Nicobar Is.

– ***glaucescens*** C. Muell. *ex* Fleisch., Musc. Fl. Buitenzorg 1: 178. 1904. ........ Andaman Is.

– ***nicobaricum*** C.Muell. *ex* Gang., Mosses E. India 2: 438. 1971. *Leucophanes nicobaricum* C. Muell., Gen. Musc. Fr. 86. 1900. ........ Nicobar Is. Endemic

– ***octoblepharoides*** Brid., Bryol. Univ. 1: 763. 1826. ........ E.H.; Nepal

***Levierella*** C. Muell., Bull. Soc. Bot. Ital. 1897: 73. 1897. (Fabroniaceae)

Type: *L. fabroniacea* C. Muell.

– ***fabroniacea*** C. Muell., Bull. Soc. Bot. Ital. 1897: 73. 1897. ........ W.H.; E.H.; S.I.; C.I.; Nepal

***Lindbergia*** Kindb., Gen. Eur. N. Am. Bryin. 15. 1897. (Leskeaceae)

Type: *L. brachypterum* {Mitt.) Kindb.

– ***duthiei*** (Broth.) Broth., Nat. Pfl. 1(3): 993. 721. 1907. *Leskea duthiei* Broth., Ac. Soc. Sc. Fenn. 24: 31. 1899. ........ W.H.

– ***koelzii*** Williams, Bryologist 35: 3. 1932. ........ W.H.; Nepal

– ***longinervis*** Card. *et* Dix., J. Bot. 50: 155-156. 1912. ........ W.H. Endemic

***Lyellia*** R.Br., Trans. Linn. Soc. London 12: 561. 1819. (Polytrichaceae)

Type: *L. crispa* R.Br.

– ***bifurcata*** Bel. *et* Bory., Voy. Ind. Orient. Bot. 2.(Crypt.) 99. 1834. ........ S.I. Endemic

– ***crispa*** R.Br., Trans. Linn. Soc. London 12: 562. 1819. ........ E.H.; Nepal

***Macrocoma*** (C.Muell.) Grout, Bryologist 47 : 4. 1944. (Orthotrichaceae)

Type: *M. filiforme* (Hook. *et* Grev.) Grout

– ***orthotrichoides*** (Raddi) Wijk *et* Marg., Taxon 11: 221. 1962. *Lasia orthotrichoides* Raddi, Critt. Bras. 6. 1822. ........ W.H.

***Macromitrium*** Brid., Mant. Musc. 132. 1819. (Orthotrichaceae)

Type: *M. aciculare* Brid.

– ***calymperoideum*** Mitt., Kew J. Bot. 8. 354. 1856. ........ E.H.; S.I.

– ***densum*** Mitt., J. Linn. Soc. Bot. Suppl. 1: 51. 1859. ........ Nepal

– ***hamatum*** Dix., J. Bombay Nat. Hist. Soc. 39: 777. 1937. ........ E.H. Endemic

– ***hymenostomum*** Mont., Ann. Sc. Nat. Bot. 4: 120. 1845. ........ W.H.; Nepal

– ***incrustatifolium*** Robins., Bryologist 71.90. 1968. ........ E.H. Endemic

– ***moorcroftii*** (Hook. *et* Grev.) Schwaegr., Spec. Musc. Suppl. 2(2): 67. 172. 1826. *Orthotrichum moorcroftii* Hook. *et* Grev., Edinburgh J. Sc. 1: 116.4. 1824. ........ W.H.; E.H.; S.I.; Nepal

– ***nepalense*** (Hook. *et* Grev.) Schwaegr., Spec. Musc. Suppl. 2(2): 134. 192. 1827. *Orthotrichum nepalense* Hook. *et* Grev., Endinburgh J. Sc. 1: 117.4. 1824. ........ E.H.; S.I.; Nepal

– ***perrottetii*** C. Muell., Syn. 1: 721. 1849. ........ E.H.; S.I.

– ***rigbyanum*** Dix., J. Bombay Nat. Hist. Soc. 39: 777. 1 *f.* 5. 1937. ........ E.H. Endemic

– ***sulcatum*** (Hook.) Brid., Bryol. Univ. 1: 319. 1826. *Schlotheimia sulcata* Hook., Musci Exot. 2: 156. 1819. ........ E.H.; S.I.; C.I.; Nepal

– ***turgidum*** Dix., J. Siam Soc. Nat. Hist. Suppl. 9: 22. 1932. ........ E.H.

***Macrothamniella*** Fleisch., Musci Fl. Buitenzorg 4: 1476. 1923. (Hylocomiaceae)

Type: *M. pilosula* (Mitt.) Fleisch.

– ***pilosula*** (Mitt.) Fleisch., Musci Fl. Buitenzorg 4: 1476. 1923. *Stereodon pilosulus* Mitt., Musci Ind. Or. 113. 1858. ........ E.H.

***Macrothamnium*** Fleisch., Hedwigia 44: 307. 1905. (Hylocomiaceae)

– ***leptohymenoides*** Nog., Kumamoto J. Sc. Biol. 11(1): 6. 1972. ........ Nepal

– ***macrocarpum*** (Reinw. *et* Hornsch.) Fleisch., Hedwigia 44: 308. 1905. *Hypnum macrocarpum* Reinw. *et* Hornsch., Nov. Act. Leop. Car. 14(2) Suppl. 725. 1829. ........ W.H.; E.H.; S.I.; Nepal

– ***pseudostriatum*** (C. Muell.) Fleisch., Hedwigia 44: 310. 1905. *Hypnum pseudo-striatum* C. Muell., Bot. Zeit. 14: 458. 1856. ........ E.H.

– ***stigmatophyllum*** Fleisch., Hedwigia 44: 310. 1905. ........ E.H.

– ***submacrocarpum*** (Ren. *et* Card.) Fleisch., Hedwigia 44: 310. 1905. *Microthamnium submacrocarpon* Jaeg. *ex* Ren. *et* Card., Bull. Soc. R. Bot. Belg. 41(1): 99. 1905. ........ W.H.; E.H.; S.I.; Nepal

***Meesia*** Hedw., Spec. Musc. 173. 1801. (Meesiaceae)

L. Type : *M. longiseta* Hedw.

– ***triquetra*** (L.) Aongstr., Nov. Act. R. Soc. Sc. Upsal. 12: 357. 1844. *Mnium triquetrum* L. *in* Richter, Codex Bot. Linn. 1045. 1840. ........ S.I.

– ***uliginosa*** Hedw., Spec. Musc. 173. 1801. ........ E.H.

***Meiothecium*** Mitt., J. Linn. Soc. Bot. 10: 185. 1868. (Sematophyllaceae)

– ***jagorii*** (C. Muell.) Broth., Nat. Pfl. 1(3): 1103. 1908. *Neckera jagorii* C. Muell., Bot. Zeit. 22: 373. 1864. ........ S.I.; Andaman Is.

– ***microcarpum*** (Hook.) Mitt., J. Linn. Soc. Bot. 10: 185. 1868. *Pterogonium microcarpum* Hook., Icon. Pl. Rar. 1: 24 f. 12. 1836. ........ E.H.; Andaman Is.; Nepal

– ***tenuirete*** Dix., Not. R. Bot. Gard. Edinburgh 19: 298. 1938. ........ Nepal

***Meteoriella*** Okam., J. Coll. Sc. Imp. Univ. Tokyo 36(7): 18. 1915. (Pterobryaceae)

Type: *M. soluta* (Mitt.) Okam.

– ***soluta*** (Mitt.) Okam., J. Coll. Sc. Imp. Univ. Tokyo 36(7): 18. 1915. *Meteorium solutum* Mitt., J. Linn. Soc. Bot. Suppl. 1: 88. 1859. ........ E.H.

***Meteoriopsis*** Fleisch. *ex* Broth., Nat. Pfl. 1(3): 825. 1906. (Meteoriaceae)

Type: *M. squarrosa* (Hook.) Fleisch.

– ***ancistrodes*** (Ren. *et* Card.) Broth., Nat. Pfl. 1(3): 826. 1906. *Meteorium ancistrodes* Ren. *et* Card., Bull. Soc. R. Bot. Belg. 34(2): 72. 1896. ........ W.H.

– ***divergens*** (Mitt.) Broth., Nat. Pfl. 1(3): 827. 1906. *Meteorium divergens* Mitt., J. Linn. Soc. Bot. Suppl. 1: 87. 1859. ........ E.H.

– ***formosana*** Nog., J. Hattori Bot. Lab. 3: 92.40. 1948.
........ W.H.; E.H.; Nepal

– ***reclinata*** (C. Muell.) Fleisch. *in* Broth., Nat. Pfl. 1(3): 826. 1906. *Pilotrichum reclinatum* C. Muell., Bot. Zeit. 12: 572. 1854.
........ W.H.; E.H.; S.I.; C.I.; Nepal

– ***squarrosa*** (Hook.) Fleisch. *in* Broth., Nat. Pfl. 1(3): 826. 1906. *Neckera squarrosa* Hook., Icon. Pl. Rar. 1: 22 *f.* 2. 1836.
........ W.H.; E.H.; S.I.; Nepal

***Meteorium*** (Brid.) Doz. *et* Molk., Musci Fr. ined. Archip. Indici 157. 1854.
(Meteoriaceae)

– ***brevirameum*** (C. Muell.) Broth., Nat. Pfl. 1(3): 818. 1906. *Neckera breviramea* C. Muell., Bot. Zeit. 12. 570. 1854. ........ S.I.

– ***buchananii*** (Brid.) Broth., Nat. Pfl. 1(3): 818. 1906. *Isothecium buchananii* Brid., Bryol.Univ., 2: 363. 1827.
........ W.H.; E.H.; S.I.; Nepal

– ***helminthocladum*** (C. Muell.) Fleisch., Musci Fl. Buitenzorg 3 : 778. 1908. *Papillaria helminthoclada* C. Muell., Nuov. Giorn. Bot. Ital. n. ser. 3: 113. 1896. ........ E.H.; Nepal

– ***miquelianum*** (C. Muell.) Fleisch. *in* Broth., Nat. Pfl. 1(3): 818. 1906. *Neckera miqueliana* C. Muell., Syn. 2: 138. 1850. ........ S.I.

***Microcampylopus*** (C. Muell.) Fleisch., Musci Fl. Buitenzorg 1: 59. 1902.
(Dicranaceae)

Type: *M. subnanus* (C. Muell.) Fleisch.

– ***subnanus*** (C. Muell.) Fleisch., Musci Fl. Buitenzorg 1: 60. 1902. *Dicranum subnanum* C. Muell., Bot. Zeit. 17: 190. 1859.
........ E.H.; S.I.

***Microdus*** Schimp. *in* Besch., Mem. Soc. Natn. Sc. Nat. Math. Cherbourh 16: 161. 1872. (Dicranaceae)

– ***assamicus*** Dix., J. Bombay Nat. Hist. Soc. 39: 773. 1937.
........ E.H. Endemic

– ***brasiliensis*** (Dub.) Ther., Bull. Herb. Boiss. ser. 2, 7: 277, 278. 1907. *Weisia brasiliensis* Dub., Mem. Soc. Phys. Hist. Nat. Geneve 7: 412. 1836. ........ W.H.; E.H.; S.I.; G.P.

***Micromitrium*** Aust., Musci Appal. 10. 1870. (Ephemeraceae)

– ***tenerum*** (B.S.G.) Crosby, Bryologist 71: 116. 1968. *Phascum tenerum* B.S.G., Bryol. Eur. (I. Mon.): 6. 1836. ........ E.H.; C.I.

***Mielichhobryum*** Srivastava, Nat. Acad. Sci. Letters 8(1): 15. 1985. (Bryaceae)

– ***sahayadrense*** (Card. *et* Dix.) Srivastava, Nat. Acad. Sci. Letters 8(1): 15. 1985. *Bryum sahayadrense* Card. *et* Dix., J. Bot. 49: 144. f.5. 1911. ........ W.H.; S.I.

***Mielichhoferia*** Nees *et* Hornsch., Bryol. Germ. 2 : 179. 1831. (Bryaceae)

– ***acutifolia*** Broth., Mitteil. Inst. Allg. Bot. Hamburg 8(2): 401. 1931. ........ W.H.

– ***assamica*** Dix., J. Bombay Nat. Hist. Soc. 39: 778. 1937. ........ E.H. Endemic

– ***badhwarii*** Dix., Ann. Bryol. 3: 60. 1930. ........ W.H.

– ***himalayana*** Mitt., J. Linn. Soc. Bot. Suppl. 1: 65. 1859. ........ W.H.

– ***lahulensis*** Williams, Bryologist 35: 1. 1. 1932. ........ W.H.

– ***mielichhoferi*** (Hook.) Wijk *et* Marg., Taxon 10: 24. 1961. *Weisia mielichhoferi* Hook., Musci Exot. 2: 201. 1819. ........ Nepal

var. ***noguchianum*** Gang., Mosses E. India 4: 893. 1974. ........ Nepal

– ***pilifera*** Bartr., Rev. Bryol. Lich. 29(3-4): 168. t 1 E-G. 1960. ........ W.H.

– ***schmidii*** C. Muell., Bot. Zeit. 11: 20. 1853. ........ S.I.

***Mitrobryum*** Robins., Bryologist 77(2): 82. 1968. (Dicranaceae)

Type: *M. koelzii* Robins.

– ***koelzii*** Robins., Bryologist 77(2): 82. 1968. ........ W.H.

***Mittenothamnium*** Henn., Hedwigia 41: 225. 1892. (Hypnaceae)

– ***molliculum*** Broth. *in* Bruehl, Rec. Bot. Surv. India 13(1): 126. 1931. ........ E.H.

***Mitthyridium*** Robins., Phytologia 32(5): 432. 1975. (Calymperaceae)

L. Type: *Syrrhopodon fasciculatus* Hook. & Grev.

- ***cardotii*** (Fleisch.) Robins., Phytologia 32: 432. 1975. *Thyridium cardotii* Fleisch., Musci Fl. Buitenzorg 1: 228. 1904. ........ E.H.
- ***fasciculatum*** (Hook. & Grev.) Robins., Phytologia 32: 433. 1975. *Syrrhopodon fasciculatus* Hook. & Grev., Edinburgh J. Sci. 3: 225. 1825. ........ S.I.; Nepal
- ***flavum*** (C. Muell.) Robins., Phytologia 32: 433. 1975. *Syrrhopodon flavus* C. Muell., Bot. Zeit. 13: 763. 1855. ........ Andaman & Nicobar Is.
- ***obtusifolium*** (Lindb.) Robins., Phytologia 32: 434. 1975. *Syrrhopodon obtusifolius* Lindb., Oefv. Svensk. Vet. Acad. Foerh. 21: 605. 1865. ........ Andaman Is.
- ***repens*** (Harv. *in* Hook.) Robins., Phytologia 32: 434. 1975. *Syrrhopodon repens* Harv. *in* Hook., Icon. Pl. Rar. 1: 22. 1836. ........ S.I.; Andaman Is.
- ***undulatum*** (Doz. *et* Molk.) Robins., Phytologia 32: 435. 1975. *Codonoblepharum undulatum* Doz. et Molk., Ann. Sc. Nat. Bot. ser. 3, 2: 301. 1844. ........ Andaman Is.

***Mniobryum*** Limpr., Laubm. Deutschl. 2 : 272. 1892. (Bryaceae)

L.Type : *M. carneum* (Web. *et* Mohr) Limpr.

- ***delicatulum*** (Hedw.) Dix., Rev. Bryol. Lich. 2: 107. 1934. *Bryum delicatulum* Hedw., Spec. Musc. 179. 1801. ........ W.H.
- ***ludwigii*** (Schwaegr.) Loesk., Stud. Morph. Syst. Laubm. : 124. 1910. *Bryum ludwigii* Spreng. *ex* Schwaegr., Spec. Musc. Suppl. 1(2): 95. 1816. ........ W.H.; E.H.
- ***wahlenbergii*** (Web. *et* Mohr) Jenn., Man. Moss W. Pensylv. 146. 1913. *Hypnum wahlenbergii* Web. *et* Mohr, Bot. Taschenb. 280, 475. 1807. ........ W.H.; E.H.

***Mnium*** Hedw., Spec. Musc. 188. 1801 (Mniaceae)

L. Type : *M. hornum* Hedw.

- ***affine*** Bland. *ex* Funck, Crypt. Gew. Ficht. 17: 3. 1810. ........ W.H.; Nepal

– ***confertidens*** (Lindb. *et* Arn.) Kindb., Enum. Bryin. Exot. 107. 1891. *Asterophyllum confertidens* Lindb. *et* Arn., K. Svensk Vet. Ak Handl. 23(10): 17. 1890. ........ E.H.; Nepal

– ***cuspidatum*** Hedw., Spec. Musc. 192, 45. f. 5-6. 1801. ........ W.H.; E.H.

ssp. ***trichomanes*** (Mitt.) Kab., Hedwigia 76: 34. 1936. *Mnium trichomanes* Mitt., Kew J. Bot. 8: 231. 1856. ........ W.H.

– ***heterophyllum*** (Hook.) Schwaegr., Spec. Musc. Suppl. 2(2): 22. 1826. *Bryum heterophyllum* Hook., Trans. Linn. Soc. 9: 318. 1808. ........ W.H.; E.H.; Nepal

– ***integrum*** Bosch *et* Lac., Bryol. Jav. 1: 153. 1861. ........ W.H.; E.H.; Nepal

– ***japonicum*** Lindb., Act. Soc. Fenn. 10 : 226. 1872. .......W.H.; E.H.; Nepal

– ***kumaonensis*** Srivastava, Nat. Acad. Sci. Letters 12(6): 177. 1989. ........ W.H.

– ***laevinerve*** Card., Bull. Soc. Bot. Geneve ser. 2, 1: 128. 1909. ........ W.H.; E.H.; Nepal

– ***lycopodioides*** Schwaegr., Spec. Musc. Suppl. 2(2): 24. 1826. ........ W.H.; E.H.; Nepal

– ***marginatum*** (With.) P. Beauv., Prodr. 75. 1805. *Bryum marginatum* Dicks. *ex* With., Syst. Arr. Brit. Pl. ed. 4. 3: 824. 1801. ........ W.H.; E.H.; Nepal

var. ***riparium*** (Mitt.) Husn., Musc. Gall. 256. 1889. *Mnium riparium* Mitt., J. Linn. Soc. Bot. 8: 30. 1864. ........ W.H.; E.H.; Nepal

– ***maximoviczii*** Lindb., Acta Soc. Sc. Fenn. 10.224. 1872. ........ W.H.; E.H.

– ***medium*** B.S.G., Bryol. Eur. 4: 196. 1838. ........ W.H.; E.H.

– ***orthorhynchum*** Brid., Bryol. Eur. 4: 25. t. 291. 1838. ........ W.H.; E.H.

– ***pseudopunctatum*** Bruch *et* Schimp., London J. Bot. 2: 669. 1843. ........ E.H.; Nepal

– ***punctatum*** Hedw., Spec. Musc. 190. 1801. ........ W.H.; E.H.; Nepal var. ***horikawae*** Nog., Trans. Nat. Hist. Soc. Formosa 24: 290. 1934. ........ W.H.; E.H.; Nepal

– ***rostratum*** Schrad., Bot. Zeit. Regensburg 1: 79. 1802. ........ W.II.; E.H.; S.I.; Nepal

– ***seligeri*** Jur., Mild. Bryol. Siles 227. 1869. ........ W.H.

– ***stellare*** Reichdt. *ex* Hedw., Spec. Musc. 191, 45. f. 1-4. 1801. ........ W.H.

– ***striatulum*** Mitt., J. Linn. Soc. Bot. ser. 2.3 : 167. 1891. ........ W.H.; E.H.

– ***succulentum*** Mitt., Musci Ind. Or. 143. 1859. ........ W.H.; E.H.; S.I.; Nepal

– ***thomsonii*** Schimp., Syn. ed. 2 : 485. 1876. ........ W.H.; E.H.; Nepal

– ***undulatum*** Weis. *ex* Hedw., Spec. Musc. 195. 1801. *Mnium serpyllifolium* var. *undulatum* L., Spec. Pl. ed. 1. 2: 1113. 1753. ........ E.H.

– ***vesicatum*** Besch., Ann. Sc. Nat. Bot. ser. 7, 17: 345. 1893. ........ E.H.

var. ***latedecurens*** Kab., Hedwigia 76: 49. 1936. ........ E.H.

***Molendoa*** Lindb., Utkast Nat. Grupp. Eur. Bladmoss. 29. 1878. (Pottiaceae)

Type: *M. hornschuchiana* (Hook.) Lindb. ex Limpr.

– ***duthiei*** (Broth.) Broth., Nat. Pfl. 1(3): 391. 1902. *Anoectangium duthiei* Broth., Act. Soc. Sc. Fenn. 24(2): 6. 1899. ........ W.H.

– ***roylei*** (Mitt.) Broth., Nat. Pfl. 1(3): 391. 1902. *Anoectangium roylei* Mitt., Musci Ind. Or. 30. 1859. ........ W.H.; Nepal

– ***sendtneriana*** (B.S.G.) Limpr., Laubm. Deutschl. 1: 250. 1886. *Anoectangium sendtnerianum* B.S.G., Bryol. Eur. 1: 91. 39. 1846 fasc. (33-36 Mon. 7.3). ........ W.H.

***Myurella*** B.S.G., Bryol. Eur. 6: 39. 1853. (Thuidiaceae)

– ***julacea*** (Schwaegr.) B.S.G., Bryol. Eur. 6: 41. 560. 1853. *Leskea julacea* Schwaegr. *in* Schultes, Reise Glockner 2: 363. 1804. ........ W.H.

- ***sibirica*** (C. Muell.) Reim., Hedwigia 76: 292. 1937. *Hypnum sibiricum* C. Muell., Syn. 2 : 418. 1851. ........ Nepal

***Myurium*** Schimp., Synop. Musc. Eur. ed. 1: 695. 1860. (Pterobryaceae)

Type: *M. hebridarum* Schimp.

- ***perplexum*** (Ren. *et* Card.) Broth., Nat. Pfl. ed. 2, 11: 124. 1925. *Lepyrodon perplexus* Ren. *et* Card., Bull. Soc. R. Bot. Belg. 34(2): 66. 1896. ........ E.H.
- ***rufescens*** (Reinw. *et* Hornsch.) Fleisch., Musci Fl. Buitenzorg 3: 672, 126. 1908. *Leucodon rufescens* Reinw. *et* Hornsch., Nov. Act. Ac. Car. Leop. Caes. 14(2): 712. 1829. ........ E.H.; S.I.
- ***warburgii*** (C. Muell.) Fleisch., Musci Archip. Indici 9: 419. 1906. *Oedicladium warburgii* C. Muell. in Broth., Monsunia 1: 176. 1900. ........ S.I.

***Nanothecium*** Dix. *et* P. Vard., Ann. Crypt. Exot. 3:181. 1930. (Hypnaceae)

Type: *N. foreaui* Dix. *et* P. Vard.

- ***foreaui*** Dix. *et* P. Vard., Ann. Crypt. Exot. 3: 181. 1930. ........ S.I.

***Neckera*** Hedw., Spec. Musc. 200. 1801. (Neckeraceae)

L. Type: *N. pennata* Hedw.

- ***aequalifolia*** C. Muell., Bot. Zeit. 12: 569. 1854. ........ S.I.
- ***andrei*** Ther. *et* P. Vard., Rev. Bryol. 50(3): 74. 1923. ........ S.I.
- ***complanata*** (Hedw.) Hueb., Musc. Germ. 576. 1833. *Leskea complanata* Hedw., Spec. Musc. 231. 1801. ........ W.H.
- ***crenulata*** Harv. *in* Hook., Icon. Pl. Rar. 1 : 21. f. 6. 1836. ........ W.H.; E.H.; Neapl
- ***goughiana*** Mitt., J. Linn. Soc. Bot. Suppl. 1: 123. 1859. ........ S.I.
- ***himalayana*** Mitt., J. Linn. Soc. Bot. Suppl. 1: 121. 1859. ........ E.H.; Nepal
- ***pennata*** Hedw., Spec. Musc. 200. 1801. ........ W.H.; E.H.; S.I.
- ***semicrispa*** Card. *et* P. Vard., Rev. Bryol. 50: 74. 1923 *et* 51: 5. 1924. ........ S.I.

– ***setschwanica*** Broth., Akad. Wiss. Wien Math. Naturw. Kl. Abt. 1. 133: 215. 1922. ........ E.H.

***Neckeropsis*** Reichdt., Reise Oesterr. Freg. Novara Bot. 1(3): 181. 1870. (Neckeraceae)

Type: *N. undulata* (Hedw.) Reichdt.

– ***acutata*** (Mitt.) Fleisch., Musci Fl. Buitenzorg 3: 878. 1907. *Neckera acutata* Mitt., J. Linn. Soc. Bot. Suppl. 1: 121. 1859. ........ E.H.

– ***andamana*** (C. Muell.) Fleisch., Musci Fl. Buitenzorg 3: 878. 1908. *Distichia andamana* C. Muell., Flora 61: 84. 1878. ........ S.I.; Andaman Is.; Nicobar Is.

– ***crinita*** (Griff.) Fleisch., Musci Fl. Buitenzorg 3: 878. 1908. *Neckera crinita* Griff., Notul. Pl. As. 2: 464. 1849 *et* Icon. Pl. As. 2: 84 f. 1. 1849. ........ E.H.; C.I.

– ***darjeelingensis*** Gang., Mosses E. India 5: 1398. 1976. ........ E.H. Endemic

– ***exserta*** (Schwaegr.) Broth., Nat. Pfl. ed. 2. 11: 188. 1925. *Neckera exserta* Hook. *ex* Schwaegr., Spec. Msuc. Suppl. 3(1): 244. 1828. ........ W.H.; E.H.; S.I.; Nepal

– ***fimbriata*** (Harv.) Fleisch., Musci Fl. Buitenzorg 3: 878. 1908. *Neckera fimbriata* Harv. *in* Hook., Icon. Pl. Rar. 1: 21 f. 4. 1836. ........ E.H.; Nepal

– ***gracilenta*** (Bosch *et* Lac.) Fleisch., Musci Fl. Buitenzorg 3: 876. 1908. *Neckera gracilenta* Bosch *et* Lac., Bryol. Jav. 2: 62. 182. 1863. ........ E.H.; Nicobar Is.

– ***lepineana*** (Mont.) Fleisch., Musci Fl. Buitenzorg 3: 879. 155. 1908. *Neckera lepineana* Mont., Ann. Sc. Nat. Bot. ser. 3, 10: 107. 1848. ........ W.H.; E.H.; S.I.

– ***submarginata*** Card. *ex* Touw, Blumea 11(2): 417. 1962. ........ S.I.

***Neobarbella*** Nog., J. Hattori Bot. Lab. 3: 72. 1948. (Meteoriaceae)

Type: *N. pilifera* (Broth. *et* Yas.) Nog.

– ***comes*** (Griff.) Nog., J. Hattori Bot. Lab. 3: 73. 1948. *Neckera comes* Griff., Calcutta Nat. Hist. 3: 71. 1843. ........ E.H.

***Neodolichomitra*** Nog., J. Jap. Bot. 13: 92. 1937. (Hylocomiaceae)

Type: *N. gigantea* Nog.

– ***robusta*** (Broth.) Nog., Bryologist 69: 232. 1966. *Penzigiella robusta* Broth., Sitzungsber. Ak. Wiss. Wien Math. Nat. Kl. 1923.
........ E.H.

***Ochrobryum*** Mitt., J. Linn. Soc. Bot. 12: 107. 1869. (Leucobryaceae)

L. Type : *O. gardneri* (C. Muell.) Lindb.

– ***nepalense*** Besch., J. de Bot. 11 : 145. 1897. ........ W.H.; Nepal

– ***propaguliferum*** Dix., Not. R. Bot. Gard. Edinburgh 19: 281, 1. 1938.
........ S.I.

***Octoblepharum*** Hedw., Spec. Musc. 50. 1801. (Calymperaceae)

Type: *O. albidum* Hedw.

– ***albidum*** Hedw., Spec. Musc. 50. 1801.
........ W.H.; E.H.; S.I.; G.P.; C.I.; Andaman Is.; Nepal

***Octogonella*** Dix., J. Bot. 74: 1. 1936. (Orthotrichaceae)

Type: *O. scabrifolia* Dix.

– ***scabrifolia*** Dix., J. Bot. 74: 1. 1936. ........ W.H.

***Okamuraea*** Broth., Oefv. Finsk. Vet. Soc. Foerh. 49: 2. 1906.
(Leskeaceae)

Type: *O. cristata* Broth.

– ***hakoniensis*** (Mitt.) Broth., Nat. Pfl. 1(3): 1133. 1908. *Hypnum hakoniense* Mitt., Trans. Linn. Soc. Bot. ser. 2. 3: 185. 1891.
........ I.O.

***Oligotrichum*** Lam. *et* Cand., Fl. France ed. 3.2: 491. 1805.
(Polytrichaceae)

– ***falcifolium*** (Griff.) G. Sm., Fl. E. Himal. 3: 244. 1975. ........ Nepal

– ***nepalense*** G. Sm., J. Hattori Bot. Lab. 41: 420. 1976. ........ Nepal

– ***obtusatum*** Broth., Symb. Sin. 4: 133. 1929. ........ E.H.

– ***semi-lamellatum*** (Hook.*f.*) Mitt., Musci Ind. Or. 150. 1859. *Polytrichum semilamellatum* Hook. *f.* in Hook., Icon. Pl. Rar. 2: t. 194 A. 1. 1837. ........ W.H.; E.H.

***Oncophorus*** (Brid.) Brid., Bryol. Univ. 1: 389. 1826. (Dicranaceae)

L. Type : *O. virens* (Hedw.) Brid.

– ***gracillimus*** Dix., Rec. Bot. Surv. India 9: 304, 1. f. 1926. ........ W.H.

– ***virens*** (Hedw.) Brid., Bryol. Univ. 1: 390. 1826. *Dicranum virens* Hedw., Spec. Musc. 142. 1801. ........ W.H.; E.H.; Nepal

– ***wahlenbergii*** Brid., Bryol. Univ. 1: 400. 1826. ........ W.H.; E.H.; Nepal

***Oreas*** Brid., Bryol. Univ. 1: 380. 1826. (Dicranaceae)

L. Type : *O. martiana* (Hopp. *et* Hornsch.) Brid.

– ***martiana*** (Hopp. *et* Hornsch.) Brid., Bryol. Univ. 1: 383. 1826. *Weisia martiana* Hopp. *et* Hornsch *in* Hook., Musc. Exot. 2: 104. 1819. ........ W.H.; E.H.

***Oreoweisia*** (B.S.G.) De Not., Epil. 1: 489. 1869. (Dicranaceae)

Type: *O. serrulata* (Funck) De Not.

– ***brevidens*** Herz., Ann. Bryol. 12: 87. 1939. ........ E.H. Endemic

– ***laxifolia*** (Hook.*f.*) Kindb., Enum. Bryol. Exot. 69. 1888. *Grimmia laxifolia* Hook.*f.* in Hook., Icon. Pl. Rar. 2: 194 B. 1837. ........ W.H.; E.H.; S.I.; Nepal

***Orontobryum*** Mitt. *ex* Fleisch., Musci Fl. Buitenzorg 4: 1475, 1478. 1922. (Hylocomiaceae)

Type: *O. hookeri* (Mitt.) Fleisch.

– ***hookeri*** (Mitt.) Fleisch. in Broth., Nat. Pfl. ed. 2, 11: 261. 1925. *Stereodon hookeri* Mitt., Musci Ind. Or. 114. 1859. ........ E.H. Endemic

– ***recurvulum*** Gang., Mosses E. India 6: 1506. 1977. ........ E.H. Endemic

***Orthodicranum*** (B.S.G.) Loesk., Stud. Morph. System. Laubm. 85. 1910. (Dicranaceae)

– ***montanum*** (Hedw.) Loesk., Stud. Morph. System Laubm. 85. 1910. *Dicranum montanum* Hedw., Spec. Musc. 143. 35 f. 8-13. 1801. ........ W.H.

***Ortholimnobium*** Dix., J. Bombay Nat. Hist. Soc. 39(4): 787. 1937. (Amblystegiaceae)

– ***borii*** Dix., J. Bombay Nat. Hist. Soc. 39(4): 788. 1937. ........ E.H. Endemic

***Orthomniopsis*** Broth., Oefv. Finsk. Vet. Soc. Foerh, 49(10): 1. 1907. (Mniaceae)

Type: *O. japonica* Broth.

– ***dilatata*** (Mitt.) Not., Flor. E. Him. Tokyo Univ. : 563. 1966. *Mnium dilatatum* Wils. *ex* Mitt., Musci Ind. Or. 143. 1859. ........ E.H.; Nepal

***Orthomnium*** (Mitt.) Broth., Oefv. Finsk. Vet. Soc. Foerh. 47(14): 7. 1905. (Mniaceae)

– ***bryoides*** (Griff.) Wijk *et* Marg., Taxon 10: 25. 1961. *Orthotrichum bryoides* Griff., Calcutta J. Nat. Hist. 2: 486. 1842. ........ W.H.; E.H.; Nepal

***Orthotheciadelphus*** Dix., J. Bot. 69: 6. 1931. (Entodontaceae)

Type: *O. ovicarpus* Dix.

– ***ovicarpus*** Dix., J. Bot. 69: 595. f. 5. 1931. ........ W.H.

***Orthothecium*** B.S.G., Bryol. Eur. 5: 105. 1851. (Hypnaceae)

Type: *O. rufescens* (Sm.) B.S.G.

– ***intricatum*** (Hartm.) B.S.G., Bryol. Eur. 5: 108, 457. 1851. *Leskea intricata* Hartm., Handb. Skand. Fl. ed. 5: 336. 1849. ........ W.H.; P.; Nepal

***Orthotrichum*** Hedw., Spec. Musc. 162. 1801. (Orthotrichaceae)

L. Type : *O. anomalum* Hedw.

– ***affine*** Schrad. *ex* Brid., Musc. Rec. 2(2): 22. 1801. ........ W.H.

var. ***fastigiatum*** (Brid.) Hueb., Musc. Germ. 365. 1833. *Orthotrichum fastigiatum* Bruch *ex* Brid., Bryol. Univ. 1: 785. 1827. ........ W.II.

– ***alpestre*** Hornsch. *in* B.S.G., Bryol. Eur. 3: 75, 213. 1849. ........ W.H.

– ***anomalum*** Hedw., Spec. Musc. 162. 1801. ........ W.H.; S.I.

– ***cupulatum*** Hoffm. *ex* Brid., Musc. Rec. 2(2): 25. 1801. ........ W.H.

– ***duthiei*** Vent., Rev. Bryol. 23: 27. 1896. ........ W.H.

– ***griffithii*** Mitt. *ex* Dix., J. Bot. 49: 140, 513, f. 2. 1911. ........ W.H.; E.H.

– ***hookeri*** Wils. *ex* Mitt., J. Linn. Soc. Bot. Suppl. 1: 48. 1859. ........ W.H.; E.H.; Nepal

– ***macounii*** Aust., Bull. Torr. Bot. Cl. 6: 343. 1879. ........ W.H.

– ***obtusifolium*** Brid., Musc. Rec. 2(2): 25. 1801. ........ W.H.

– ***pumilum*** Sw., Monthl. Rev. 34: 538. 9. 1801. ........ W.H.

– ***rupestre*** Schleich. *ex* Schwaegr., Spec. Musc. Suppl. 1(2): 27. 53. 1816. ........ W.H.

– ***sikkimense*** Herz., Ann. Bryol. 12: 90. 11. 1939. ........ E.H. Endemic

– ***speciosum*** Nees *in* Sturm., Deutschl. Fl. 2(3) (fasc. 17): 5. 1819. ........ W.H.; E.H.

– ***striatum*** Hedw., Spec. Musc. 163. 1801. ........ W.H.

– ***sturmii*** Hopp. *et* Hornsch, Bot. Zeit. 2(6): 89. 1819. ........ W.H.

– ***urnigerum*** Myr., Coroll. Fl. Upsal. 71. 1833. ........ W.H.

– ***venustum*** Vent. *in* Broth., Acta Soc. Sc. Fenn. 24(2): 18. 1898. ........ W.H.

– ***virens*** Vent. *in* Broth., Acta Soc. Sc. Fenn. 24(2): 18. 1898. ........ W.H.; S.I.

***Osterwaldiella*** Fleisch. *ex* Broth., Nat. Pfl. ed. 2. 11: 130. 1925. (Pterobryaceae)

Type: *O. monostricta* Broth.

– ***monostricta*** Fleisch. *ex* Broth., Nat. Pfl. ed. 2. 11: 131. 1925. ........ E.H.

***Oxystegus*** (Limpr.) Hilp., Beih. Bot. Centralbl. 50(2): 666, 667. 1933. (Pottiaceae)

– ***burmense*** (Bartr.) Gang., Mosses E. India 3: 653. 1972. *Trichostomum burmense* Bartr., Rev. Bryol. Lich. 23: 245. 1954. ........ I.O.

– ***cylindricus*** (Brid.) Hilp., Beih. Bot. Centralbl. 50(2): 667. 1933. *Weissia cylindrica* Bruch *ex* Brid., Bryol. Univ. 1: 806. 1827. .......W.H.; E.H.; S.I.; Nepal

– ***cylindrothecus*** (Mitt.) Gang., Nov. Hedwigia 12: 430. 1966. *Tortula cylindrotheca* Mitt., Musci Ind. Or. 30. 1859. ........ E.H. Endemic

– ***indicus*** (Dix. *et* P. Vard.) Hilp., Beih. Bot. Zbl. 50(2): 670. 1926. ........ S.I.

– ***khasianus*** (Mitt.) Gang., Nov. Hedwigia 8: 149. 1964. *Tortula khasiana* Mitt., Musci Ind. Or. 29. 1859. ........ E.H. Endemic

– ***stenophyllus*** (Mitt.) Gang., Nov. Hedwigia 12: 429. 1966. *Tortula stenophylla* Mitt., Musci Ind. Or. 28. 1859. ........ W.H.; E.H.; S.I.; Endemic

– ***tenuirostris*** (Hook. & Tayl.) A.J.E. Sm., J. Bryol. 9: 393. 1977. *Weisia tenuirostris* Hook. & Tayl., Muscol. Brit. Ed. 2. 83, Suppl. Tab. 3. 1827. ........ W.H.; E.H.; S.I.; Nepal

var. ***gemmiparus*** (Schimp.) Zand., Lindbergia 4: 285. 1978. *Didymodon cylindricus* var. *gemmiparus* Schimp., Synop. Musc., Ed. 2. 165. 1876. ........ W.H.

***Palamocladium*** (C. Muell.) Robins., Bryologist 65(2): 95. 1962. (Brachytheciaceae)

– ***euchloron*** (C. Muell.) Wijk *et* Marg., Taxon 9: 52. 1960. *Hypnum euchloron* Bruch *ex* C. Muell., Syn. 2: 464. 1851. ........ W.H.

– ***nilgheriense*** (Mont.) C. Muell., Flora 82: 465. 1896. *Isothecium nilgheriense* Mont., Ann. Sc. Nat. Bot. ser. 2. 17: 246. 1842. ........ W.H.; E.H.; S.I.

***Papillaria*** (C. Muell.) C. Muell., Oefv. K. Svensk. Vet. Ak. Foerh. 33(4): 34. 1876. (Meteoriaceae)

– ***chrysoclada*** (C. Muell.) Jaeg., Ber. S. Gall. Naturw. Ges. 1875-76: 270. 1877. *Neckera chrysoclada* C. Muell., Syn. 2: 139. 1850. ........ E.H.; S.I.

– ***crocea*** (Hamp.) Jaeg., Ber. S. Gall. Naturw. Ges. 1875-76: 267. 1877. ........ S.I.

– ***formosana*** Nog.

var. ***pilifera*** Nog., Kumamoto J. Sc. ser. B. Sect. 2, 4: 3. 1958. ........ E.H.; S.I.

– ***fuscescens*** (Hook.) Jaeg., Ber. S. Gall. Naturw. Ges. 1875-76 : 270. 1877. *Neckera fuscescens* Hook., Musci Exot. 2: 157. 1819. ........ E.H.; S.I.; Nepal

– ***semitorta*** (C. Muell.) Jaeg., Ber. S. Gall. Naturw. Ges. 1875-76: 271. 1877. *Neckera semitorta* C. Muell., Syn. 2: 671. 1851. ........ W.H.; E.H.; S.I.; Nepal

***Paraleucobryum*** (Limpr.) Loesk., Allg. Bot. Zeit. 13: 167. 1907. (Dicranaceae)

Type: *P. albicans* (Schwaegr.) Loesk.

– ***enerve*** (Thed.) Loesk., Hedwigia 47: 171. 1908. *Dicranum enerve* Thed. in Hartm., Handb. Skand. Fl. ed. 5: 393. 1849. ........ W.H.; E.H.; Nepal

– ***himalayanum*** Dix., 150 th Anniv. R. Bot. Gard. Calcutta 177. 1942. ........ W.H.

***Pelekium*** Mitt., J. Linn. Soc. Bot. 10: 176. 1868. (Thuidiaceae)

Type: *P. velatum* Mitt.

– ***bifarium*** (Bosch *et* Lac.) Fleisch., Musci Fl. Buitenzorg 4: 1513. 1923. *Thuidium bifarium* Bosch *et* Lac., Bryol. Jav. 2: 123. 226. 1865. ........ E.H.

– ***velatum*** Mitt., J. Linn. Soc. Bot. 10: 176. 1868. ........ E.H.

***Penzigiella*** Fleisch., Hedwigia 45: 86. 1906. (Pterobryaceae)

Type: *P. cordata* (Harv.) Fleisch.

- ***cordata*** (Harv.) Fleisch., Hedwigia 45: 87. 1906. *Hypnum cordatum* Harv. *in* Hook., Icon. Pl. Rar. 1: 24 *f.* 7. 1836. ........ W.H.; E.H.; S.I.; Nepal

***Philonotis*** Brid., Bryol. Univ. 2: 15. 1827. (Bartramiaceae)

L. Type : *P. fontana* (Hedw.) Brid.

- ***alpicola*** Jur. *ex* Limpr., Laubm. Deutschl. 11: 573. 1893. ........ S.I.
- ***angusta*** Mitt., Musci Ind. Or. 61. 1859. ........ W.H.; E.H.
- ***anisoclada*** Card. *et* P. Vard., Rev. Bryol. 50 (2): 23. 1923. ........ S.I.
- ***calcarea*** (B.S.G.) Schimp., Coroll. Bryol. Eur. 86. 1856. *Barıramia calcarea* B.S.G., Bryol. Eur. 4: 49. 325. 1842. ........ W.H.
- ***falcata*** (Hook.) Mitt., Musci Ind. Or. 62. 1859. *Bartramia falcata* Hook., Trans. Linn. Soc. London 9: 317. 1808. ........ W.H.; E.H.; S.I.; G.P.; Nepal
- ***fontana*** (Hedw.) Brid., Bryol. Univ. 2 : 18. 1827. *Mnium fontanum* L. *ex* Hedw., Spec. Musc. 195. 1801. ........ W.H.; E.H.; S.I.; Nepal

  var. ***pumilla*** (Turn.) Brid., Bryol. Univ. 2: 20. 1827. *Bartramia fontana* var. *pumila* Turn., Musc. Hib. 107. 10 f. 1. 1804. ........ W.H.
- ***glomerata*** Mitt., Musci Ind. Or. 60. 1859. ........ E.H.; S.I.
- ***hastata*** (Dub.) Wijk *et* Marg., Taxon 8: 74. 1959. *Hypnum hastatum* Dub. *in* Moritzi, Syst. Verz. Zoll. Pfl. 132. 1846. ........ W.H.; E.H.; G.P.; S.I.
- ***heterophylla*** Mitt., J. Linn. Soc. Bot. Suppl. 1: 61. 1859. ........ W.H.; S.I.
- ***lancifolia*** Mitt., J. Linn. Soc. Bot. 8: 151. 1865. ........ W.H.; Nepal
- ***leptocarpa*** Mitt., Musci Ind. Or. 60. 1859. ........ W.H.; E.H.
- ***marchica*** (Hedw.) Brid., Bryol. Univ. 2: 23. 735. 1827. *Mnium marchicum* Hedw., Spec. Musc. 196. 1801. ........ Himalayas

– ***mollis*** (Doz. *et* Molk.) Mitt., Musci Ind. Or. 60. 1859. *Bartramia mollis* Doz. *et* Molk., Ann. Sc. Nat. Bot. ser. 3.2: 300. 1844. ........ E.H.; S.I.; Andaman Is.

***nitida*** Mitt., Musci Ind. Or. 62. 1859. ........ W.H.; E.H.

– ***revoluta*** Bosch *et* Lac., Bryol. Jav. 1: 158, 128. 1861. ........ E.H.

– ***rigida*** Brid., Bryol. Univ. 2: 17. 1827. ........ S.I.

– ***secunda*** (Doz. *et* Molk.) Bosch *et* Lac., Bryol. Jav. 1: 156, 126. 1861. *Bartramia secunda* Doz. et Molk., Plant. Jungh. 3: 332. 1854. ........ S.I.

– ***seriata*** Mitt., Musci Ind. Or. 63. 1859. ........ W.H.; Nepal

– ***socia*** Mitt., J. Linn. Soc. Bot. 8: 151.1. 1865. ........ Nepal

– ***speciosa*** (Griff.) Mitt., Musci Ind. Or. 64. 1859. *Bartramia speciosa* Griff., Calcutta J. Nat. Hist. 2: 513. 1842. ........ E.H.; Nepal

– ***subrigida*** Card. *et* P. Vard., Rev. Bryol. 50(2): 12. 1923. ........ S.I.

– ***tomentella*** Mol. *in* Lor., Moosstud. 170. 1864. ........ S.I.

– ***trachyphylla*** Dix. *et* Badhw., Rec. Bot. Surv. India 12: 175. 1938. ........ W.H.

– ***turneriana*** (Schwaegr.) Mitt., Musci Ind. Or. 62. 1859. *Bartramia turneriana* Schwaegr., Spec. Musc. Suppl. 3(1): 238. 1828. ........ W.H.; E.H.; Nepal

– ***thwaitesii*** Mitt., Musci Ind. Or. 60. 1859. ........ W.H.; E.H.; S.I.; C.I.; Nicobar Is.; Nepal

***Physcomitrium*** (Brid.) Brid., Flora 13 (2 Erg.) : 9. 1829. (Funariaceae)

L. Type : *P. sphaericum* (Ludw.) Fuernr.

– ***coorgense*** Broth., Rec. Bot. Surv. India 1: 319. 1899. ........ E.H.; S.I.; G.P.; C.I.

– ***cyathicarpum*** Mitt., Musci Ind. Or. 54. 1859. ........ W.H.; C.I.; G.P.; P.; R.; Nepal

– ***eurystomum*** Sendtn., Denkschr. Bayer Bot. Ges. Regensburg 3: 142. 1841. ........ W.H.; E.H.; R.; G.P.

– ***indicum*** (Dix.) Gang., Bull. Bot. Soc. Bengal 23: 131. 1969. *Physcomitrellopsis indica* Dix. *in* Gupta, J. Indian Bot. Soc. 13: 122. 1933. ........ G.P.

– ***insigne*** Dix. *et* P. Vard., Arch. Bot. 1(8-9): 168, 3 f.5. 1927. ........ S.I.

– ***japonicum*** (Hedw.) Mitt., Trans. Linn. Soc. Bot. London ser. 2, 3: 164. 1891. *Gymnostomum japonicum* Hedw., Spec. Musc. 34. 1801. ........ W.H.; E.H.; G.P.

– ***pulchellum*** (Griff.) Mitt., Musci Ind. Or. 54. 1859. *Gymnostomum pulchellum* Griff., Calcutta J. Nat. Hist. 2: 478. 1842. ........ E.H.; G.P.

– ***repondum*** (Griff.) Mitt., Musci Ind. Or. 54. 1859. *Gymnostomum repondum* Griff., Calcutta J. Nat. Hist. 2: 418. 1842. ........ E.H.; P.; Nepal

***Pilotrichopsis*** Besch., J. Bot. Paris 13(2): 38. 1899. (Cryphaeaceae)

Type: *P. dentata* (Mitt.) Besch.

– ***dentata*** (Mitt.) Besch., J. Bot. Paris 13(2): 38. 1899. *Dendropogon dentatus* Mitt., Trans. Linn. Soc. Bot. ser. 2, 3: 170. 1891. ........ E.H.

– ***ferruginea*** (Mitt.) Broth., Nat. Pfl. ed. 2.11: 86. 1925. *Cryphaea ferruginea* Mitt., J. Linn. Soc. Bot. Suppl. 1: 125. 1859. ........ S.I.

***Pinnatella*** Fleisch., Hedwigia 45: 79. 1906. (Neckeraceae)

– ***alopecuroides*** (Hook.) Fleisch., Hedwigia 45: 84. 1906. *Hypnum alopecuroides* Hook., Icon. Pl. Rar. 1: 24 *f.* 5. 1836. ........ E.H.; S.I.; Nepal

– ***ambigua*** (Bosch *et* Lac.) Fleisch., Hedwigia 45: 81. 1906. *Thamnium ambiguum* Bosch *et* Lac., Bryol. Jav. 2: 72. 192. 1863 .......W.H.; E.H.; C.I.

– ***anacamptolepis*** (C. Muell.) Broth., Nat. Pfl. 1(3): 857. 1906. *Neckera anacamptolepis* C. Muell., Syn. 2: 633. 1851. ........ S.I.

– ***calcuttensis*** Fleisch., Hedwigia 45: 84. 1906. ........ S.I.; C.I.; G.P.

– ***foreauna*** Ther. *et* P. Vard., Rev. Bryol. 52: 39. 1925. ........ S.I.

– ***gollani*** Broth., Mitteil. Inst. Allg. Bot. Hamburg 8(2): 404. 1931. ........ W.H.

– ***kurzii*** (Kindb.) Wijk *et* Marg., Taxon 10: 222. 1961. *Thamnium kurzii* Kindb., Hedwigia 41: 246. 1902. ........ E.H.; Nepal

– ***limbata*** Dix., J. Indian Bot. 2: 184, 8. 1921. ........ S.I.

– ***micropter**a* Fleisch., Musci Fl. Buitenzorg 3: 915. 1908. ........ E.H.

– ***sikkimensis*** Broth., Mitteil. Inst. Allg. Bot. Hamburg 8: 404. 1931. ........ E.H. Endemic

***Plagiobryum*** Lindb., Ocfv. K. Vct. Ak. Foerh. 19 : 606. 1863. (Bryaceae)

L. Type : *P. zierii* (Hedw.) Lindb.

– ***demissum*** (Hook.) Lindb., Oefv. K. Vet. Ak. Foerh. 19 : 606. 1863. *Bryum demissum* (Hornsch.) Hook., Musci Exot. 2: 16.99. 1819. ........ E.H.

– ***zierii*** (Hedw.) Lindb., Oefv K. Vet. Ak. Foerh. 19: 606. 1863. *Bryum zierii* Dicks. *ex* Hedw., Spec. Musc. 182. 1801. ........ E.H.

***Plagiomnium*** T. Kop., Ann. Bot. Fenn. 5: 145. 1968. (Mniaceae)

Type : *P. cuspidatum* (Hedw.) T. Kop.

– ***rostratum*** (Schrad.) T. Kop. ........ W.H.

***Plagiopus*** Brid., Bryol. Univ. 1: 596. 1826. (Bartramiaceae)

Type: *P. serratus* Brid.

– ***oederi*** (Brid.) Limpr., Laubm. Deutschl. 2: 548. 1895. *Bartramia oederi* Brid., Musc. Rec. 2(3): 135. 1803. ........ W.H.

***Plagiothecium*** B.S.G., Bryol. Eur. 5: 179. 1851. (Plagiotheciaceae)

L. Type : *P. denticulatum* (Hedw.) B.S.G.

– ***cavifolium*** (Brid.) Iwats., J. Hattori Bot. Lab. 33: 360. 1970. *Hypnum cavifolium* Brid., Bryol. Univ. 2: 556. 1827. ........ Nepal

– ***cochleatum*** Dix., Not. R. Bot. Gard. Edinburgh 19: 299, 12. 1938. ........ W.H.

– ***denticulatum*** (Hedw.) B.S.G., Bryol. Eur. 5: 190.501. 1851. *Hypnum denticulatum* L. *ex* Hedw., Spec. Musc. 237. 1801. ........ W.H.; E.H.; S.I.; Nepal

– ***entodontella*** Broth. *ex* Dix., J. Siam. Soc. Nat. Hist. Suppl. 10(1): 28. 1935. ........ E.H. Endemic

– ***neckeroideum*** B.S.G., Bryol. Eur. 5: 195. 1851. ........ W.H.; E.H; S.I.; Nepal

var. ***sikkimense*** Ren. *et* Card., Bull. Soc. R. Bot. Belg. 41(1): 108. 1905. ........ W.H.; E.H; S.I.; Nepal

- ***nemorale*** (Mitt.) Jaeg., Ber. S. Gall. Naturw. Ges. 1876-77: 451. 1878. *Stereodon nemoralis* Mitt., Musci Ind. Or. 104. 1859. ........ W.H.; E.H.; Nepal
- ***paleaceum*** (Mitt.) Jaeg., Ber. S. Gall. Naturw. Ges. 1876-77: 452. 1878. *Stereodon paleaceus* Mitt., Musci Ind. Or. 103. 1859. ........ E.H. Endemic
- ***perminutum*** Dix., Ann. Bryol. 3: 67. 1930. ........ W.H.
- ***roeseanum*** B.S.G., Bryol. Eur. 5: 193, 504. 1851. ........ W.H.; Nepal
- ***sylvaticum*** (Brid.) B.S.G., Bryol. Eur. 5: 192.503. 1851. *Hypnum sylvaticum* Brid., Musc. Rec. 2(2): 53. 1 *f.* 5. 1801. ........ W.H.; E.H.; P.
- ***vesiculariopsis*** Dix. *et* P. Vard., Arch. Bot. 1(8-9): 174.8.*f*.1. 1927. ........ S.I.

***Platydictya*** Berk., Handb. Brit. Moss. 145. 1863. (Hypnaceae)

Type: *P. sprucei* (B.S.G.) Berk.

- ***jungermannioides*** (Brid.) R.S. Chopra, Taxon. Indian Mosses: 449. 1975. *Hypnum jungermannioides* Brid., Muscol. Recent. Suppl. 2: 255. 1812. ........ W.H.
- ***madurensis*** (Card. & P. Vard.) R.S. Chopra, Taxon. Indian Mosses: 450. 1975. *Amblystegiella madurensis* Card. & P. Vard., Rev. Bryol. 50(3): 78. 1923. ........ S.I.
- ***subtilis*** (Hedw.) Crum, Mich. Bot. 3: 60. 1964. *Leskea subtilis* Hedw., Spec. Musc. 221. 1801. ........ W.H.

***Platygyrium*** B.S.G., Bryol. Eur. 5: 95. 1851. (Hypnaceae)

Type: *P. repens* (Brid.) B.S.G.

- ***repens*** (Brid.) B.S.G., Bryol. Eur. 5: 98, 458. 1851. *Pterigynandrum repens* Brid., Spec. Musc. 1: 131. 1806. ........ W.H.
- ***russulum*** (Mitt.) Jaeg., Ber. S. Gall. Naturw. Ges. 1876-77: 276. 1878. *Stereodon russulus* Mitt., Musci Ind. Or. 94. 1859. ........ W.H.; E.H.; Nepal
- ***subrussulum*** Ren. *et* Card., Bull. Soc. R. Bot. Belg. 38(1): 37. 1900. ........ E.H. Endemic

***Pleuridiella*** Robins., J. Hattori Bot. Lab. 27: 125. 1964. (Ditrichaceae)

– ***colei*** Robins., J. Hattori Bot. Lab. 27: 125. 1964. ........ E.H.

***Pleuridium*** Brid., Mant. Musc. 4: 10. 1819. (Ditrichaceae)

– ***denticulatum*** (C. Muell.) Mitt., J. Linn. Soc. Bot. Suppl. 1: 7. 1859. *Astomum denticulatum* C. Muell., Bot. Zeit. 11: 18. 1853. ........ S.I.

– ***mussuriense*** Broth. *ex* Roth., Aussereur. Laubm. 153. 16. f. 5. 1911. ........ W.H.

– ***tenue*** Mitt., Musci Ind. Or. 7. 1859. *Phascum tenue* Wils., Kew J. Bot. 9: 290. 1857. *nom. nud.* ........ W.H.; E.H.

***Pleurochaete*** Lindb., Oefv. K. Vet. Ak. Foerh. 21: 253. 1864. (Pottiaceae)

Type: *P. squarrosa* (Brid.) Lindb.

– ***squarrosa*** (Brid.) Lindb., Oefv. K. Vet. Ak. Foerh. 21: 253. 1864. *Barbula squarrosa* Brid., Bryol. Univ. 1: 833. 1827. ........ W.H.; S.I.

***Pleurozium*** Mitt., J. Linn. Soc. Bot. 12: 537. 1869. (Hylocomiaceae)

L. Type : *P. schreberi* (Brid.) Mitt.

– ***schreberi*** (Brid.). Mitt., J. Linn. Soc. Bot. 12: 537. 1869. *Hypnum schreberi* Willd. *ex* Brid., Musc. Rec. 2(2): 88. 1801. ........ E.H.; Nepal

***Pogonatum*** P. Beauv., Mag. Enc. 5: 329. 1804. (Polytrichaceae)

– ***akitense*** Besch., Ann. Sc. Nat. Bot. ser. 7. 17: 354. 1893. ........ E.H.

– ***aloides*** (Hedw.) P. Beauv., Prodr. 84. 1805. *Polytrichum aloides* Hedw., Spec. Musc. 96. 1801. ........ W.H.; E.H.; S.I.; Nepal

– ***alpinum*** (Hedw.) Roehl., Ann. Wetterau Ges. 3(2): 226. 1814. .......W.H.; E.H.; Nepal

– ***cirratum*** (Sw.) Brid., Bryol. Univ. 2: 110. 1827. *Polytrichum cirratum* Sw., J.F. Bot. 1800(2): 176. 1802. ........ E.H.; Nepal

– ***contortum*** (Brid.) Lesq., Mem. Calif. Ac. Sc. 1: 27. 1868. *Polytrichum contortum* Menz. *ex* Brid., J.F. Bot. 1800(2): 287. 1801. ........ E.H.; Nepal

– ***decolyi*** Gang., Nov. Hedwigia 12: 420. 1966. ........ E.H.; Nepal

– ***fastigiatum*** Mitt., Musci Ind. Or. 154. 1859. ........ E.H.; Nepal

– ***flexicaule*** Mitt., Musci Ind. Or. 152. 1859. ........ E.H.

– ***fuscatum*** Mitt., Musci Ind. Or. 154. 1859. ........ W.H.; E.H.; Nepal

– ***grandifolium*** (Lindb.) Jaeg., Ber. S. Gall. Naturw. Ges. 1873-74: 253. 1875. *Polytrichum grandifolium* Lindb., Act. Soc. Sc. Fenn. 10: 264. 1872. ........ E.H.

– ***gymnophyllum*** Mitt., Musci Ind. Or. 153. 1859. ........ E.H.

– ***hexagonum*** Mitt., Musci Ind. Or. 151. 1859. ........ W.H.; E.H.; S.I.; Nepal

– ***himalayanum*** Mitt., Musci Ind. Or. 151. 1859. ........ W.H.; E.H.; S.I.; Nepal

– ***inflexum*** (Lindb.) Lac., Ann. Mus. Bot. Lugd. -Bot. 4: 308. 1869. *Polytrichum inflexum* Lindb., Not. Saellsk. F. Fl. Fenn. Foerh. 9: 100. 1868. ........ S.I.

– ***junghuhinianum*** (Doz. *et* Molk.) Doz. *et* Molk., Bryol. Jav. 1: 41. 1856. *Polytrichum junghuhinianum* Doz. *et* Molk., Pl. Jungh. 3: 324. 1854. ........ W.H.; E.H.; S.I.; Nepal

var. ***sikkimense*** Ren. *et* Card., Bull. Soc. R. Bot. Belg. 34: 65. 1895. ........ W.H.; E.H.; S.I.; Nepal

– ***leucopogon*** Ren. *et* Card., Bull. R. Bot. Soc. Belg. 34: 64. 1896. ........ E.H. Endemic

– ***macrophyllum*** Doz. *et* Molk., Bryol. Jav. 1: 45. 35. 1856. ........ E.H.

– ***microstomum*** (Schwaegr.) Brid., Bryol. Univ. 2: 745. 1827. *Polytrichum microstomum* R.Br. *ex* Schwaegr., Spec. Musc. Suppl. 2(2): 10. 1826. ........ W.H.; E.H.; S.I.; Nepal

– ***muticum*** Broth., Sitzungsber. Ak. Wiss. Wien. Math. Nat. Kl. Abt. 133 : 583. 1924. ........ Nepal

– ***norkettii*** R.S. Chopra, Taxon. Indian Mosses: 25. 1975. ........ E.H.

– ***nudiusculum*** Mitt., Musci Ind. Or. 153. 1859. ........ E.H.; Nepal

– ***papillosulum*** Card. *et* Dix., J. Bot. 50: 146. 1912. ........ E.H. Endemic

– ***patulum*** (Harv.) Mitt., Musci Ind. Or. 152. 1859. *Polytrichum patulum* Harv. *in* Hook., Icon. Pl. Rar. 1: 18 f. 1. 1836.
........ Nepal

– ***perichaetiale*** (Mont.) Jaeg., Ber. S. Gall. Naturw. Ges. 1873-74: 257. 1875. *Polytrichum perichaetiale* Mont., Ann. Sc. Nat. Bot. ser. 2. 17: 252. 1842. ........ W.H.; E.H.; S.I.; Nepal

– ***proliferum*** (Griff.) Mitt., Musci Ind. Or. 152. 1859. *Polytrichum proliferum* Griff., Calcutta J. Nat. Hist. 2(8): 475. 1842.
........ E.H.; Nepal

– ***rufisetum*** Mitt., Musci Ind. Or. 153. 1859. ........ E.H.; Nepal

– ***seminudum*** Mitt., Musci Ind. Or. 152. 1859.
........ W.H.; E.H.; Nepal

– ***stevensii*** Ren. *et* Card., Bull. R. Bot. Soc. Belg. 34: 65. 1895.
........ W.H.; E.H.; Nepal

– ***strictifolium*** Gang., Nov. Hedwigia 12: 419. 1966.
........ E.H. Endemic

– ***subtortile*** (C.Muell.) Jaeg., Ber. S. Gall. Naturw. Ges. 1873-74: 256. 1875. *Polytrichum subtortile* C. Muell., Syn. 1: 216. 1848.
........ E.H.; S.I.

var. ***teysmannianum*** (Doz. *et* Molk.) Wijk et Marg. Taxon. 9: 191. 1960. *Polytrichum teysmannianum* Doz. & Molk., Pl. Jungh. 3: 323. 1854. ........ E.H.; S.I.

– ***thomsonii*** (Mitt.) Jaeg., Ber. S. Gall. Naturw. Ges. 1873-74: 257. 1875. *Polytrichum thomsonii* Mitt., J. Linn. Soc. Bot. Suppl. 1: 155. 1859. ........ W.H.

– ***tortipes*** (Mitt.) Jaeg., Ber. S. Gall. Naturw. Ges. 1873-74: 257. 1875. *Polytrichum tortipes* Wils. *ex* Mitt., Musci Ind. Or. 155. 1859.
........ E.H.; Nepal

– ***urnigerum*** (Hedw.) P. Beauv., Prodr. 84. 1805. *Polytrichum urnigerum* L. *ex* Hedw., Spec. Musc. 100. 1801.
........ W.H.; E.H.; Nepal

***Pohlia*** Hedw., Spec. Musc. 171. 1801. (Bryaceae)

Type: *P. elongata* Hedw.

– ***ampullacea*** Gang., Mosses E. India 4: 921. 1974.
........ E.H. Endemic

– ***camptotrachela*** (Ren. *et* Card.) Broth., Nat. Pfl 1(3): 552. 1903. *Webera camptotrachela* Ren. *et* Card., Bot. Gaz. 13: 199. 1888. .......W.H.; E.H.; S.I.; Nepal

– ***cruda*** (Hedw.) Lindb., Musci Scand. 18. 1879. *Mnium crudum* L. *ex* Hedw., Spec. Musc. 189. 1801. ........ W.H.; E.H.; Nepal

– ***crudoides*** (Sull. *et* Lesq.) Broth., Nat. Pfl. 1(3): 548. 1903. *Bryum crudoides* Sull. *et* Lesq., Proc. Am. Ac. Arts Sc. 4: 278.1859. ........ W.H.; Nepal

– ***elongata*** Hedw., Spec. Musc. 171. 1801. ........ W.H.; E.H.; S.I.; Nepal

– ***flexuosa*** Hook., Icon. Pl. Rar. 1: 19. 1836. ........ W.H.; E.H.; S.I.; G.P.; Nepal

var. ***propagulifera*** (Ren. *et* Card.) Gang., Mosses E. India 4: 910. 1974. ........ E.H.

– ***gedeana*** (Bosch *et* Lac.) Gang., Mosses E. India 4: 927. 1974. *Bryum gedeanum* Bosch *et* Lac., Bryol. Jav. 1: 147. 1860. ........ I.O.

– ***himalayana*** (Mitt.) Broth., Nat. Pfl. 1(3): 548. 1903. *Webera himalayana* Mitt., Musci Ind. Or. 66. 1859. ........ E.H. Endemic

– ***longicollis*** (Hedw.) Lindb., Musci Scand. 18. 1879. *Webera longicolla* Hedw., Spec. Musc. 169. 1801. ........ W.H.; E.H.; Nepal

– ***minor*** Schleich. *ex* Schwaegr., Spec. Musc. Suppl. 1(2): 70. 1816. .......W.H.; E.H.

ssp. ***acuminata*** (Hopp. *et* Hornsch.) Wijk *et* Marg., Taxon 8: 74. 1959. *Pohlia acuminata* Hopp. *et* Hornsch., Flora 2: 94. 1819. .......W.H.; E.H.; Nepal

– ***rigescens*** (Mitt.) Broth., Nat. Pfl. 1(3): 548. 1903. *Webera rigescens* Mitt., Musci Ind. Or. 65. 1859. ........ E.H. Endemic

***Polytrichastrum*** G. Sm., Mem. N.Y. Bot. Gard. 21(3): 35. 1971. (Polytrichaceae)

Type: *Polytrichum alpinum* L. ex Hedw.

– ***emodi*** G. Sm., J.Hattori Bot. Lab. 38: 633. 1974. ........ Nepal

– ***papillatum*** G. Sm., J. Hattori Bot. Lab. 38: 633. 1974. ........ W.H.; Nepal

– ***torquatum*** Osada & G. Sm., J. Hattori Bot. Lab. 38: 636. 1974. .......Nepal

– ***xanthopilum*** (Wils. ex Mitt.) G. Sm., Mem. N.Y. Bot. Gard. 21(3): 35. 1971. *Polytrichum xanthopilum* Wils. ex Mitt., J. Linn. Soc. Bot. Suppl. 1: 156. 1859. ........ E.H.

***Polytrichum*** Hedw., Spec. Musc. 88. 1801. (Polytrichaceae)

Type: *P. commune* L. ex Hedw.

– ***alpinum*** L. *ex* Hedw., Spec. Musc. 92. 1801. ........ W.H.; E.H.; Nepal

– ***densifolium*** Wils. *ex* Mitt., Musci Ind. Or. 155. 1859. ........ E.H.

– ***formosum*** Hedw., Spec. Musc. 92. 1801. ........ Nepal

– ***juniperinum*** Willd. *ex* Hedw., Spec. Musc. 89. 18 f. 6-10. 1801. ........ W.H.

– ***piliferum*** Schreb. *ex* Hedw., Spec. Musc. 90. 1801. ........ W.H.

***Porotrichum*** (Brid.) Hamp., Linnaea 32: 154. 1863. (Neckeraceae)

L. Type : *P. longirostre* (Hook.) Mitt.

– ***fruticosum*** (Mitt.) Jaeg., Ber. S. Gall. Naturw. Ges. 1875-76: 306. 1877. *Neckera fruticosa* Mitt., J. Linn. Soc. Bot. Suppl. 1: 122. 1859. ........ W.H.; S.I.

***Pottia*** (Reichenb.) Fuernr., Flora 12 (Erg. 2): 10. 1829. (Pottiaceae)

L. Type : *P. truncata* (Hedw.) B.S.G.

– ***alpicola*** Dix., Ann. Bryol. 3: 57. 1930. ........ W.H.

– ***starckeana*** (Hedw.) C. Muell., Syn. 1: 547. 1849. *Weisia starckeana* Hedw., Spec. Musc. 65. 1801. ........ W.H.

– ***watsonii*** R.S. Chopra, Taxon. Indian Mosses 155. 1975. ........ S.I.

***Prionidium*** Hilp., Beih. Bot. Centralbl. 50(2): 640. 1933. (Pottiaceae)

Type: *P. setschwanicum* (Broth.) Hilp.

– ***setschwanicum*** (Broth.) Hilp., Beih. Bot. Centralbl. 50(2): 641. 1933. *Leptodontium setschwanicum* Broth., Sitzungsber. Ak. Wiss. Wien Math. Nat. Kl. Abt. 1. 131: 211. 1922. ........ Nepal

***Pseudobarbella*** Nog., J. Hattori Bot. Lab. 2: 81. 1947. (Meteoriaceae)

Type: *P. assimilis* (Card.) Nog.

– ***compressiramea*** (Ren. *et* Card.) Nog., Musci Fl. East. Himalayas 3: 574. 1966. *Meteorium compressirameum* Ren. *et* Card., Bull. Soc. R. Bot. Belg. 38(1): 27. 1900. ........ E.H.; Nepal

– ***levieri*** (Ren. *et* Card.) Nog., J. Hattori Bot. Lab. 3: 86, 36. 1948. *Meteorium levieri* Ren. *et* Card., Bull. Soc. R. Bot. Belg. 41(1): 78. 1905. ........ E.H.; Nepal

***Pseudocrossidium*** Williams, Bull. Torr. Bot. Cl. 42: 396. 1915. (Pottiaceae)

Type: *P. chilense* Williams

– ***porphyreoneurum*** (C. Muell. *ex* Vent.) Zand., Bull. Buffalo Soc. Nat. Sc. NY 32: 119. 1993. *Barbula porphyreoneura* C. Muell. ex Vent. in Nuovo Giorn. Bot. Ital. 4: 13. 1872. ........ R.

***Pseudoleskea*** B.S.G., Bryol. Eur. 5: 147. 1852. (Leskeaceae)

L. Type : *P. atrovirens* (Brid.) B.S.G.

– ***incurvata*** (Hedw.) Loesk., Hedwigia 50: 313. 1911. *Leskea incurvata* Hedw., Spec. Musc. 216. 53f. 8-14. 1801. ........ E.H.; Nepal

– ***laevifolia*** (Mitt.) Jaeg., Ber. S. Gall. Naturw. Ges. 1876-77: 243. 1878. *Leskea laevifolia* Mitt., J. Linn. Soc. Bot. Suppl. 1: 130. 1859. ........ W.H. Endemic

– ***ramuligera*** (Mitt.) Sauerb. *et* Jaeg., Ber. S. Gall. Naturw. Ges. 1877-78: 475. 1880. *Leskea ramuligera* Mitt., J. Linn. Soc. Bot. Suppl. 1: 132. 1859. ........ E.H.; Nepal

***Pseudoleskeella*** Kindb., Gen. Eur. N. Am. Bryin. 20. 1897. (Leskeaceae)

– ***catenulata*** (Schrad.) Kindb., Gen. Eur. N. Am. Bryin. 1: 48. 1897. *Pterigynandrum catenulatum* Brid. *ex* Schrad., J. Bot. 1801: 195. 1803. ........ W.H.

***Pseudoleskeopsis*** Broth., Nat. Pfl. 1(3): 1002. 1907. (Leskeaceae)

– ***decurvata*** (Mitt.) Broth., Nat. Pfl. 1(3): 1003. 1907. *Leskea decurvata* Mitt., J. Linn. Soc. Bot. 8: 154. 1864. ........ W.H.; E.H.

– ***orbiculata*** (Mitt.) Broth., Nat. Pfl. 1(3): 1003. 1907. *Hypnum orbiculatum* Mitt., J. Linn. Soc. Bot. Suppl. 1: 84. 1859. ........ E.H.; S.I.

– ***serrulata*** Card. *et* Ther., Bull. Ac. Int. Geogr. Bot. 21: 271. 1911. ........ W.H.

– ***zippelii*** (Doz. *et* Molk.) Broth., Nat. Pfl. 1(3): 1003. 1907. *Hypnum zippelii* Doz. *et* Molk., Ann. Sc. Nat. Bot. ser. 3, 2: 310. 1844. .......W.H.; E.H.; S.I.

***Pseudospiridentopsis*** (Broth.) Fleisch., Musci Fl. Buitenzorg 3: 729. 1908. (Trachypodaceae)

Type: *P. horrida* (Card.) Fleisch.

– ***horrida*** (Card.) Fleisch., Musci Fl. Buitenzorg 3: 730. 1908. *Meteorium horridum* Mitt. *ex* Card., Beih. Bot. Centralbl. 19(2): 118. 17. 1905. ........ E.H.

***Pseudostereodon*** (Broth.) Fleisch., Musci Fl. Buitenzorg 4: 1376. 1923. (Hypnaceae)

Type: *P. procerrimum* (Mol.) Fleisch.

– ***procerrimum*** (Mol.) Fleisch., Nat. Pfl. ed. 2. 11: 455. 1925. *Hypnum procerrimum* Mol., Flora 49: 458. 1866. ........ W.H.

***Pseudosymblepharis*** Broth., Nat. Pfl. ed. 2. 10: 261. 1924. (Pottiaceae)

– ***angustata*** (Mitt.) Hilp., Beih. Bot. Centralbl. 50(2): 670. 1933. *Tortula angustata* Mitt., J. Proc. Linn. Soc. Bot. Suppl. 1: 28. 1859. ........ W.H.; S.I.

– ***duriuscula*** (Mitt.) Chen, Hedwigia 80: 153, 15 f. 11. 1941. *Tortula duriuscula* Mitt., J. Linn. Soc. Bot. Suppl. 1: 27. 1859. ........ S.I.

– ***pallidens*** Dix., J. Bombay Nat. Hist. Soc. 39: 776. 1 *f.* 3. 1937. ........ E.H. Endemic

– ***subduriuscula*** (C. Muell.) Chen, Hedwigia 80: 152. 15 f. 8-10. 1940. *Barbula subduriuscula* C. Muell., Linnaea 38: 554. 1874. ........ W.H.; Nepal

***Pterigynandrum*** Hedw., Spec. Musc. 80. 1801. (Pterigynandraceae)

L. Type : *P. filiforme* Hedw.

– ***decolor*** (Mitt.) Broth., Nat. Pfl. 1(3): 892. 1907. *Stereodon decolor* Mitt., Musci Ind. Or. 92. 1859. ........ E.H. Endemic

– ***filiforme*** Hedw., Spec. Musc. 81. 1801. ........ W.H.

***Pterobryopsis*** Fleisch., Hedwigia 45: 56. 1905. (Pterobryaceae)

L. Type : *P. crassicaulis* (C. Muell.) Fleisch.

– ***acuminata*** (Hook.) Fleisch., Hedwigia 45: 59. 1905. *Neckera acuminata* Hook., Musci Exot. 2: 15. 151. 1819. ........ E.H.; S.I.; Nepal

– ***auriculata*** Dix., J. Bombay Nat. Hist. Soc. 39: 782. 1 f. 10. 1937. ........ E.H.; S.I.; Nepal

– ***conchophylla*** (Ren. *et* Card.) Broth., Nat. Pfl. 1(3): 803. 1906. *Garovaglia conchophylla* Ren. *et* Card., Bull. Soc. R. Bot. Belg. 41(1): 69. 1905. ........ E.H. Endemic

– ***divergens*** (Mitt.) Nog., J. Hattori Bot. Lab. 28: 147. 1965. *Meteorium divergens* Mitt., J. Linn. Soc. Bot. Suppl. 1: 87. 1859. ........ E.H.; Nepal

– ***flexipes*** (Mitt.) Fleisch., Hedwigia 45: 62. 1905. *Meteorium flexipes* Mitt., J. Linn. Soc. Bot. Suppl. 1: 85. 1859. ........ E.H.; S.I.; Nepal

– ***nematosa*** (C. Muell.) Dix., J. Siam Soc. Nat. Hist. Suppl. 10: 13. 1935. *Meteorium nematosum* C.Muell., Flora 61.84. 1878. ........ E.H.; Nepal

– ***orientalis*** (C. Muell.) Fleisch., Hedwigia 59: 217. 1917. *Neckera orientalis* C. Muell., Bot. Zeit. 14: 437. 1856. ........ W.H.; E.H.; S.I.; Nepal

– ***scabriuscula*** (Mitt.) Fleisch., Hedwigia 45: 61. 1905. *Meteorium seabriusculum* Mitt., Proc. Linn. Soc. Suppl. 1: 85. 1859. ........

– ***tumida*** (Hook.) Dix., J. Bot. 75: 122. 1937. *Neckera tumida* Dicks. *ex* Hook., Musci Exot. 2: 158. 1819. ........ Nepal

***Ptilium*** De Not. in Cronac., Briol. Ital. 2: 17. 1867. (Hypnaceae)

Type: *P. crista-castrensis* (Hedw.) De Not.

– ***crista-castrensis*** (Hedw.) De Not. in Cronac., Briol. Ital. 2: 17. 1867. *Hypnum crista-castrens* L. *ex* Hedw., Spec. Musc. 287. 1801. ........ E.H.; Nepal

***Ptychomitrium*** Fuernr., Flora 12 Erg. 2: 19. 1829. (Ptychomitriaceae)

Type: *P. polyphyllum* (Sw.) B.S.G.

– ***indicum*** (Schrad.) Jaeg., Ber. S. Gall. Naturw. Ges. 1872-73: 104. 1874. *Trichostomum indicum* Willd. *ex* Schrad., J. Bot. 1801(1): 196. 1803. ........ E.H.

– ***fauriei*** Besch., J. de Bot. Paris 12: 297. 1898. ........ E.H.

– ***tortula*** (Harv. *in* Hook.) Jaeg., Ber. S. Gall. Naturw. Ges. 1872-73: 105. 1874. *Didymodon tortula* Harv. *in* Hook., Icon. Pl. Rar. 1: 18 f. 2.1836. ........ W.H.; E.H.; S.I.

***Pylaisia*** B.S.G., Bryol. Eur. 5: 87. 1851. (Hypnaceae)

– ***brevirostris*** (Griff.) Jaeg., Ber. S. Gall. Naturw. Ges. 1877-78 : 481. 1880. *Neckera brevirostris* Griff., Calcutta J. Nat. Hist. 3: 69. 1843. ........ E.H. Endemic

– ***extenta*** (Mitt.) Jaeg., Ber. S. Gall. Naturw. Ges. 1876-77: 306. 1878. *Stereodon extentus* Mitt., J. Linn. Soc. Bot. Suppl. 1: 95. 1859. ........ E.H.; Nepal

– ***falcata*** B.S.G., Bryol. Eur. 5: 89. 1851. ........ E.H.

– ***polyantha*** (Hedw.) B.S.G., Bryol. Eur. 5: 88, 445. 1851. *Leskea polyantha* Hedw., Spec. Musc. 229. 1801. ........ W.H.

***Pylaisiadelpha*** Card., Rev. Bryol. 39: 57. 1912. (Sematophyllaceae)

Type: *P. rhaphidostegioides* (Card.) Card.

– ***drepanioides*** Card. *et* Dix., Rev. Bryol. 39: 58. 1912. ........ W.H.; E.H.

***Pylaisiopsis*** Broth., Nat. Pfl. 1(3): 1232. 619. 1909. (Sematophyllaceae)

– ***speciosa*** (Mitt.) Broth., Nat. Pfl. 1(3): 1232. 619. 1909. *Stereodon speciosus* Mitt., Musci Ind. Or. 96. 1859. ........ E.H. Endemic

***Racomitrium*** Brid., Mant. Musc. 78. 1819. (Grimmiaceae)

– ***canescens*** (Hedw.) Brid., Mant. Musc. 78. 1819. *Trichostomum canescens* Hedw., Spec. Musc. 111. 1801. ........ W.H.; E.H.

– ***crispulum*** (Hook. *f. et* Wils.) Hook. *f. et* Wils., Fl. Nov. Zel. 2: 75. 1854. *Dryptodon crispulus* Hook. *f. et* Wils., London J. Bot. 3: 544. 1844. ........ W.H.; E.H.; S.I.; Nepal

– ***fasciculare*** (Hedw.) Brid., Mant. Musc. 78. 1819. *Trichostomum fasciculare* Schrad. ex Hedw., Spec. Musc. 110. 1801. ........ Nepal

– ***fuscescens*** Wils., Kew J. Bot. 9: 324. 1857. ........ W.H.; E.H.

– ***heterostichum*** (Hedw.) Brid., Mant. Musc. 79. 1819. *Trichostomum heterostichum* Hedw., Spec. Musc. 109. 1801. ........ E.H.; Nepal

– ***himalayanum*** (Mitt.) Jaeg., Ber. S. Gall. Naturw. Ges. 1872-73: 97. 1874. *Grimmia himalayana* Mitt., Musci Ind. Or. 45. 1859. ........ W.H.; E.H.; Nepal

– ***lanuginosum*** (Hedw.) Brid., Mant. Musc. 79. 1819. *Trichostomum lanuginosum* Hedw., Spec. Musc. 109. 1801. ........ E.H.

– ***strictifolium*** (Mitt.) Jaeg., Ber. S. Gall. Naturw. Ges. 1872-73: 89. 1874. *Grimmia strictifolia* Mitt., Musci Ind. Or. 44. 1859. ........ E.H. Endemic

– ***subsecundum*** (Hook. *et* Grev.) Mitt., Kew J. Bot. 9: 324. 1857. *Trichostomum subsecundum* Hook. *et* Grev., Icon. Pl. Rar. 1: t. 17. 1836. ........ W.H.; E.H.; S.I.; Nepal

***Racopilum*** P. Beauv., Prodr. 36. 1805. (Racopilaceae)

L. Type : *R. mnioides* P. Beauv.

– ***cuspidigerum*** (Schwaegr.) Aongstr., Oefv. K. Vet. Ak. Foerh. 29(4): 10. 1872. *Hypnum cuspidigerum* Schwaegr. *in* Gaud. *in* Freyc., Voyage Aut. Monde Oranie Phys. Bot. 229. 1828. ........ S.I.; Nepal

– ***orthocarpum*** Wils. *ex* Mitt., J. Linn. Soc. Bot. Suppl. 1: 136. 1859. ........ W.H.; E.H.; S.I.; Nepal

– ***schmidii*** (C. Muell.) Mitt., J. Linn. Soc. Bot. Suppl. 1: 136. 1859. *Hypopterygium schmidii* C. Muell., Bot. Zeit. 12: 558. 1858. ........ S.I.

***Regmatodon*** Brid., Bryol. Univ. 2: 204. 1827. (Regmatodontaceae)

Type: *R. declinatus* (Hook.) Brid.

– ***declinatus*** (Hook.) Brid., Bryol. Univ. 2 : 204. 1827 . *Pterogonium declinatum* Hook., Trans. Linn. Soc. London 9: 309.26 *f.* 3. 1808. .......W.H.; E.H.; Nepal

– ***orthostegius*** Mont., Ann. Sc. Nat. Bot. ser. 2. 17 : 248. 1842. ....... W.H.; E.H.; S.I.; Nepal

***Reimersia*** Chen, Hedwigia 80: 62. 1941. (Pottiaceae)

Type: *R. inconspicua* (Griff.) Chen

– ***inconspicua*** (Griff.) Chen, Hedwigia 80: 62. 1940. *Gymnostomum inconspicuum* Griff., Calcutta J. Nat. Hist. 2: 480. 1842. ....... W.H.; E.H.

***Retidens*** Dix., J. Bot. 74: 8. 1936. (Entodontaceae)

Type: *R. stewartii* Dix.

– ***stewartii*** Dix., J. Bot. 74: 8. 9. 1936. ....... W.H.

***Rhabdoweisia*** B.S.G., Bryol. Eur. 1 : 97. 1846. (Rhabdoweisiaceae)

– ***crenulata*** (Mitt.) Jameson, Rev. Bryol. 17: 6. 1890. *Didymodon crenulatus* Mitt., Musc. Ind. Or. 23. 1859. ....... W.H.; E.H.

– ***crispata*** (With.) Lindb., Act. Soc. Sc. Fenn. 10: 22. 1871. *Bryum crispatum* Dicks. *ex* With., Syst. Arr. Brit. Pl. ed. 4, 3: 816. 1801. ....... E.H.

***Rhachithecium*** Broth. *ex* Le Jolis, Mem. Soc. Sc. Nat. Math. Cherbourg 29: 308. 1895. (Rhachitheciaceae)

– ***perpusillum*** (Thwait. *et* Mitt.) Broth., Nat. Pfl. 1(3): 1199. 1909. *Zygodon perpusillus* (Thwait. *et* Mitt.) Broth., J. Linn. Soc. Bot. 13: 303. 1873. ....... E.H.; S.I.; Nepal

***Rhamphidium*** Mitt., J. Linn. Soc. Bot. 12(1): 45. 1869. (Pottiaceae)

Type: *R. macrostegium* (Sull.) Mitt.

– ***cribbifolium*** Dix., 150 th Ann. Vol. R. Bot. Gard. Calcutta 179. 1942. ....... W.H.

– ***laticuspis*** Dix., 150 th Ann. Vol. R. Bot. Gard. Calcutta 179. 1942. ....... W.H.

– ***madurense*** Dix. *et* P. Vard., Arch. de Bot. 1(8-9) 165, 3 *f.* 3. 1929. ....... S.I.

– ***mussuriensis*** Dix., 150 th Ann. Vol. R. Bot. Gard. Calcutta 179. 1942. ....... W.H.

***Rhaphidorrhynchium*** Fleisch., Musci Fl. Buitenzorg 4: 1245. 1923. (Sematophyllaceae)

Type: *R. aurescens* (Jaeg.) Fleisch.

– ***confertissimum*** (Mitt.) Broth., Nat. Pfl. ed. 2, 11: 424. 1925. *Stereodon confertissimus* Mitt., Musci Ind. Or. 102. 1859. ....... E.H. Endemic

***Rhaphidostichum*** Fleisch., Musci Fl. Buitenzorg 4: 1307. 1923. (Sematophyllaceae)

– ***camptocladum*** Card. *in* Foreau, J. Bombay Nat. Hist. Soc. 58(1): 43. 1961. ....... S.I.

– ***cucullifolium*** (Card. *et* Dix.) Broth., Nat. Pfl. ed. 2, 11: 434. 1925. *Sematophyllum cucullifolium* Card. *et* Dix., Rec. Bot. Surv. India 6: 88. 2 f. 6. 1914. ....... S.I.

– ***glauco-virens*** (Mitt.) Broth., Nat. Pfl. ed. 2, 11: 435. 1925. *Stereodon glauco-virens* Mitt., J. Linn. Soc. Bot. Suppl. 1: 103. 1859. ....... E.H.

– ***luxurians*** (Doz. *et* Molk.) Fleisch., Musci Fl. Buitenzorg 4: 1310. 211. 1923. *Hypnum luxurians* Doz. *et* Molk., Ann. Sc. Nat. Bot. ser. 3, 2: 307. 1844. ....... E.H.; Nicobar Is.

– ***subleptocarpum*** (Ther. *et* P. Vard.) Broth., Nat. Pfl. ed. 2, 11: 434. 1925. *Sematophyllum subleptocarpum* Ther. *et* P. Vard., Rev. Bryol. 51: 11. 17. 1924. ....... S.I.

***Rhizogonium*** Brid., Bryol. Uuiv. 2: 644. 1827. (Rhizogoniaceae)

– ***spiniforme*** (Hedw.) Bruch *in* Krauss., Flora 29: 134. 1846. *Hypnum spiniforme* L. *ex* Hedw., Spec. Musc. 236. 1801. ....... W.H.; E.H.; S.I.; Nepal

***Rhodobryum*** (Schimp.) Hamp., Linnaea 38: 663. 1874. (Bryaceae)

Type: *R. roseum* (Hedw.) Limpr.

- ***giganteum*** (Schwaegr.) Par., Ind. Bryol.: 1116. 1898. *Mnium giganteum* Schwaegr., Spec. Musc. Suppl. 2(2): 20. 158. 1826. ....... W.H.; E.H.; S.I.; Nepal

- ***laxe-limbatum*** (Ochi) Iwats. *et* Kop., Act. Bot. Fenn. 96 : 14. 1972. *Rhodobryum laxe-limbatum* (Ochi) Gang. in Singh, Calcutta Univ. Ph.D. Thesis. 1971. ined. ....... E.H.; Nepal

- ***madurense*** Dix. *et* P. Vard., Ann. Crypt. Exot. 3: 177.3 *f.* 1. 1930. ....... S.I.

- ***ontariense*** (Kindb.) Par. *in* Kindb., Eur. N. Am. Bryin 2: 346. 1897. *Bryum ontariense* Kindb., Bull. Torr. Bot. Cl. 16: 96. 1889. ....... W.H.; Nepal

- ***roseum*** (Hedw.) Limpr., Laubm. Deutschl. 3: 444. 1892. *Mnium roseum* Hedw., Spec. Musc. 194. 1801. ....... W.H.; E.H.; Nepal

***Rhynchostegiella*** (B.S.G.) Limpr., Laubm. Deutschl. 3: 207. 1896. (Brachytheciaceae)

Type: *R. tenella* (Dicks.) Limpr.

- ***assamica*** Card. *et* Dix., Rec. Bot. Surv. India 6: 72. 2. *f.* 7. 1914. ....... E.H. Endemic

- ***compacta*** (Hook.) Loesk., Stud. Morph. Syst. Laubm. 182. 1910. ....... W.H.

- ***divaricatifolia*** (Ren. *et* Card.) Broth., Nat. Pfl. 1(3): 1161. 1909. *Rhynchostegium divaricatifolium* Ren. *et* Card., Bull. Soc. R. Bot. Belg. 41(1): 91. 1905. ....... E.H. Endemic

- ***fabroniadelphus*** (C. Muell.) Broth., Nat. Pfl. l(3): 1161. 1909. *Hypnum fabroniadelphus* C. Muell., Linnaea 36: 15. 1869. ....... S.I.

- ***humillima*** (Mitt.) Broth., Nat. Pfl. 1(3): 1161. 1909. *Hypnum humillimum* Mitt., J. Linn. Soc. Bot. Suppl. 1: 80. 1859. ....... W.H.; E.H.; S.I.

- ***leiopoda*** Dix. *ex* P. Vard., Arch. Bot. 1(8-9): 172. 8 *f.* 5. 1927. ....... S.I.

- ***menadensis*** (Lac.) Bartr., Philipp. J. Sc. 68: 302. 385. 1939. *Hypnum menadense* Lac., Bryol. Jav. 2: 156. 255. 1866. ....... E.H.

– ***percomplanata*** Dix., J. Bombay Nat. Hist. Soc. 39: 788. 1937. ....... E.H.

– ***ramicola*** (Broth.) Broth., Nat. Pfl. 1(3): 1161. 1901. *Rhynchostegium ramicola* Broth. *in* Ren. *et* Card., Bull. Soc. R. Bot. Belg. 41(1): 91. 1905. ....... E.H.

– ***sachensis*** Dix., Ann. Bryol. 3: 70. 1930. ....... W.H. Endemic

– ***scabriseta*** (Schwaegr.) Broth., Nat. Pfl. 1(3): 1161. 1909. *Hypnum scabrisetum* Schwaegr., Spec. Musc. Suppl. 3(2): 281b. 1830. ....... W.H.; E.H.; Nepal

***Rhynchostegium*** B.S.G., Bryol. Eur. 5: 197. 1852. (Brachytheciaceae)

Type: *R. confertum* (Dicks.) B.S.G.

– ***calderii*** Vohra, Bull. Bot. Surv. India 22: 124. 1980. ....... E.H. Endemic

– ***celebicum*** (Lac.) Jaeg., Ber. S. Gall. Naturw. Ges. 1876-77: 374. 1878. *Hypnum celebicum* Lac., Bryol. Jav. 2 : 159.258. 1866. .......W.H.; E.H.; S.I.

– ***duthiei*** C. Muell. *ex* Dix., J. Bombay Nat. Hist. Soc. 39: 789. 1937. ....... W.H.; S.I.; E.H.

– ***herbaceum*** (Mitt.) Jaeg., Ber. S. Gall. Naturw. Ges. 1876-77 : 368. 1878. *Hypnum herbaceum* Mitt., J. Linn. Soc. Bot. Suppl. 1: 81. 1859. ....... W.H.; E.H.; S.I.

– ***hookeri*** Jaeg., Ber. S. Gall. Naturw. Ges. 1876-77: 369. 1878. ....... W.H.; E.H.

– ***muelleri*** Jaeg., Ber. S. Gall. Naturw. Ges. 1876-77: 378. 1878. ....... W.H.; E.H.; S.I.

– ***pellucidum*** Dix., J. Bombay Nat. Hist. Soc. 39. 789. 1937. ....... E.H. Endemic

– ***planiusculum*** (Mitt.) Jaeg., Ber. S. Gall. Naturw. Ges. 1876-77: 364. 1878. *Hypnum planiusculum* Mitt., J. Linn. Soc. Bot. Suppl. 1: 81. 1859. ....... W.H.; E.H.

– ***riparioides*** (Hedw.) Card. *in* Tourret, Bull. Soc. Bot. France 60: 231. 1913. *Hypnum riparioides* Hedw., Spec. Musc. 242. 1801. ....... W.H.; E.H.; S.I.; Nepal

– ***subrectocarpum*** (Dix.) Vohra, Bull. Bot. Surv. India 21: 171. 1979. *Brachythecium subrectocarpum* Dix., Rec. Bot. Surv. India 12: 177. 1938. ....... W.H. Endemic

– ***vagans*** (Harv.) Jaeg., Ber. S. Gall. Naturw. Ges. 1876-77: 369. 1878. *Hypnum vagans* Harv. *in* Hook., Icon. Pl. Rar. 1. 24 *f.* 2. 1836. ....... W.H.; E.H.; S.I.; Nepal

***Rhytidiadelphus*** (Limpr.) Warnst., Krypt. Fl. Brandenburg 2: 842, 917. 1906. (Hylocomiaceae)

– ***triquetrus*** (Hedw.) Warnst., Krypt. Fl. Brandenburg 920: 996. 926 *f.* 6. 1906. *Hypnum triquetrum* L. *ex* Hedw., Spec. Musc. 256. 1801. ....... W.H.; E.H.

***Rhytidium*** (Sull.) Kindb., Bih. K. Svensk. Vet. Ak. Handl. 6(19): 8. 1882. (Rhytidiaceae)

Type: *R. rugosum* (Hedw.) Kindb.

– ***rugosum*** (Hedw.) Kindb., Bih. K. Svensk. Vet. Ak. Handl. 7(9): 15. 1883. *Hypnum rugosum* Ehrh. *ex* Hedw., Spec. Musc. 293. 1801. ....... W.H.; Nepal

***Rozea*** Besch., Mem. Soc. Sc. Nat. Cherbourg 16: 241. 1872. (Brachytheciaceae)

– ***pterogonioides*** (Harv.) Jaeg., Ber. S. Gall. Naturw. Ges. 1876-77: 274. 1878. *Leskea pterogonioides* Harv. *in* Hook., Icon. Pl. Rar. 1: 24 *f.* 8. 1836. ....... W.H.; E.H.; Nepal

***Saelania*** Lindb., Utkast Nat. Grupp. Eur. Bladmoss. 35. 1878. (Ditrichaceae)

Type: *S. caesia* (P. Beauv.) Lindb.

– ***glaucescens*** (Hedw.) Broth. *in* Bomanss *et* Broth., Herb. Mus. Fenn. 2: 53. 1894. ....... W.H.

***Schistidium*** Brid., Mont. Musc. 20. 1819. (Grimmiaceae)

L. Type : *S. pulvinatum* (Hedw.) Brid.

– ***alpicola*** (Hedw.) Limpr., Laubm. Deutschl. 1(1): 707. 1889. *Grimmia alpicola* Sw. *ex* Hedw., Spec. Musc. 77. 15 *f.* 1-5. 1801. .......W.H.

– ***apocarpum*** (Hedw.) B.S.G. ssp. ***gracile*** (Roehl.) Meyl., Bull. Soc. Vaudoise Sc. Nat. ser. 5. 41 : 100. 1905. *Grimmia apocarpa* Hedw. var. *gracile* Roehl., Deutschl. Fl. Krypt. ed. 2, 3: 47. 1813. ....... W.H.; E.H.

***Schlotheimia*** Brid., Spec. Musc. 2: 16. 1812. (Orthotrichaceae)

L. Type : *S. torquata* (Hedw.) Brid.

– ***grevilleana*** Mitt., J. Linn. Soc. Bot. Suppl. 1: 53. 1859. ....... E.H.; S.I.

***Schoenobryum*** Doz. *et* Molk., Musci Fr. Ined. Archip. Indici 6: 183. 1848. (Cryphaeaceae)

Type: *S. julaceum* Doz. *et* Molk.

– ***concavifolium*** (Griff.) Gang., Mosses E. India 5: 1209. 1976. *Orthotrichum concavifolium* Griff., Calcutta J. Nat. Hist. 2: 400. 1842. ....... E.H.; S.I.; Nepal

– ***julaceum*** Doz. *et* Molk., Musci Fr. Ined. Archip. Indici 6: 184. 60. 1848. ....... E.H.; S.I.; Nepal

***Schwetschkea*** C. Muell., Linnaea 39: 429. 1875. (Leskeaceae)

Type: *S. schweinfurthii* C. Muell.

– ***applanata*** (Thwait. *et* Mitt.) Broth., Nat. Pfl. 1(3): 908. 1907. *Hypnum applanatum* Thwait. *et* Mitt., J. Linn. Soc. Bot. 13: 321. 1873. ....... S.I.

– ***indica*** Broth., Rec. Bot. Surv. India 1: 328. 1899. ....... S.I.

***Schwetschkeopsis*** Broth., Nat. Pfl. 1(3): 877. 1907. (Myriniaceae)

– ***fabronia*** (Schwaegr.) Broth., Nat. Pfl. 1(3): 878. 1907. *Helicodontium fabronia* Schwaegr., Spec. Musc. Suppl. 3(2): 294. 1830. ....... Nepal

– ***formosana*** Nog., J. Hattori Bot. Lab. 5: 41. 2. 1951. ....... E.H.; Nepal

– ***neckeroides*** (Griff.) Vohra, Rec. Bot. Surv. India 23: 19. 1983. *Pterogonium neckeroides* Griff., Calcutta J. Nat. Hist. 3: 64. 1843. ....... E.H. Endemic

***Sciaromium*** (Mitt.) Mitt., J. Linn. Soc. Bot. 12: 571. 1869. (Amblystegiaceae)

L. Type : *S. conspissatum* (Hook. *f. et* Wils.) Mitt.

– ***sikkimense*** Par., Ind. Bryol. Suppl. 305. 1900. ....... E.H. Endemic

***Scopelophila*** (Mitt.) Lindb., Acta Soc. Sci. Fenn. 10: 269. 1872. (Pottiaceae)

– ***cataractae*** (Mitt.) Broth., Nat. Pfl. 1(3): 436. 1902. *Weisia cataractae* Mitt., J. Linn. Soc. Bot. 12: 135. 1869. ....... W.H.

– ***ligulata*** (Spruc.) Spruc., J. Bot. n. ser. 10: 14. 1881. *Encalypta ligulata* Spruc., Musci Pyren. no. 331. 1847. ....... W.H.

***Sematophyllum*** Mitt., J. Linn. Soc. Bot. 8: 5. 1864. (Sematophyllaceae)

Type: *S. demissum* (Wils.) Mitt.

– ***angusticuspis*** Broth., Rec. Bot. Surv. India 1: 316. 1899. ....... S.I.

– ***caespitosum*** (Hedw.) Mitt., J. Linn. Soc. Bot. 12: 479. 1869. *Leskea caespitosa* Hedw., Spec. Musc. 233, 49 f. 1-5. 1801. ....... E.H.; S.I.; Nepal

– ***humile*** (Mitt.) Broth., Nat. Pfl. ed. 2. 11: 431. 1925. *Stereodon humilis* Mitt., Musci Ind. Or. 102. 1859. ....... W.H.; Nepal

– ***micans*** (Mitt.) Braithw., Brit. Moss. Fl. 3: 154. 113 B. 1902. *Stereodon micans* Mitt., Musci Ind. Or. 114. 1859. ....... E.H.

– ***phoeniceum*** (C. Muell.) Fleisch., Musci Fl. Buitenzorg 3: 1266. 1923. *Hypnum phoeniceum* C. Muell., Flora 61: 85. 1878. ....... S.I.; Andaman Is.

– ***sebillei*** (Broth. *et* Ther.) R.S. Chopra, Taxon. Indian Mosses: 512. 1975. *Rhaphidostegium sebillei* Broth. *et* Ther., Rev. Bryol. 51(1): 10. 18. 1924. ....... S.I.

– ***subcylindricum*** (Fleisch.) Sainsb., Rev. Bryol. Lich. 21: 224. 1952. *Rhaphidostegium subcylindricum* Broth. *ex* Fleisch., Hedwigia 44: 323. 1905. ....... S.I.

– ***subhumile*** (C. Muell.) Fleisch., Musci Fl. Buitenzorg 4: 1264, 205. 1923. *Hypnum subhumile* C. Muell., Syn 2: 330. 1851. ....... E.H.; S.I.; Nepal

***Semibarbula*** Herz. *ex* Hilp., Beih. Bot. Centralbl. 50(2): 626. 1933. (Pottiaceae)

- ***orientalis*** (Web.) Wijk *et* Marg., Taxon 8: 75. 1959. *Trichostomum orientale* Web., Arch. Syst. Nat. 1. 129. 1804. ....... W.H.; E.H.; G.P.; C.I.; Nicobar Is.; Nepal
- ***ranui*** Gang., Nov. Hedwigia 8: 149. 1964. ....... C.I.; G.P.; Nepal

***Solmsiella*** C. Muell., Bot. Centralbl. 19: 149. 1884. (Erpodiaceae)

- ***biseriata*** (Aust.) Steere, Bryologist 37: 100. 1934. *Lejeunea biseriata* Aust., Proc. Acad. Nat. Sci. Phila. 1869: 225. 1869. ....... E.H.
- ***ceylonica*** (Thwait. *et* Mitt.) C. Muell., Beih. Bot. Centralbl. 19: 149. 1884. *Erpodium ceylonicum* Thwait. et Mitt., J. Linn. Soc. Bot. 13: 306. 5A. 1873. ....... S.I.

***Sphaerotheciella*** Fleisch., Hedwigia 55: 282. 1914. (Cryphaeaceae)

Type: *S. sphaerocarpa* (Hook.) Fleisch.

- ***sphaerocarpa*** (Hook.) Fleisch., Hedwigia 55: 282. 1914. *Neckera sphaerocarpa* Hook., Trans. Linn. Soc. 9: 312. 27 f. 1. 1808. ....... E.H.; Nepal

***Sphagnum*** L., Spec. Plant. 1106. 1753. (Sphagnaceae)

L. Type : *S. palustre* L.

- ***acutifolioides*** Warnst., Hedwigia 29: 192, 4 f. 4; 7 f. 16. 1890. ....... E.H.
- ***beccarii*** Hamp., Nuov. Giorn. Bot. Ital. 4: 278. 1872. ....... E.H.
- ***ceylonicum*** Mitt. *ex* Warnst., Hedwigia 29: 195, 4. f. 7; f. 14. 1890. ....... S.I.
- ***cuspidatulum*** C. Muell., Linnaea 38: 549. 1874. ....... E.H.; Nepal
- ***cuspidatum*** Ehrh. *ex* Hafm., Deutschl. Fl. 2: 22. 1796. ....... E.H.
- ***fimbriatum*** Wils. *in* Hook., Crypt. Bot. Antarct. Voyage 92. 1845. ....... W.H.
- ***girgensohnii*** Russ., Arch. Naturk. Livl. Ehstl. Kurl. ser. 2, 7: 124. 1865. ....... W.H.

- ***imbricatum*** Hornsch. *ex* Russ., Arch. Naturk. Livl. Ehstl. Kurl. ser. 2, 7: 99. 1865. ....... E.H.
- ***junghuhnianum*** Doz. *et* Molk., Nat. Verb. K. Ak. Wet. Amsterdam 2: 8. 1854. ....... E.H.; Nepal
- ***khasianum*** Mitt., Musci Ind. Or. 156. 1859. ....... E.H.
- ***magellanicum*** Brid., Musc. Rec. 2(1): 24. 1798. ....... E.H.
- ***nemoreum*** Scop., Fl. Carn. ed. 2: 305. 1772. ....... E.H.
- ***nepalense*** H. Suzuki, Flora East Himalaya: 540. 1966. ....... Nepal
- ***ovatum*** Hamp. *in* C. Muell., Linnaea 38: 546. 1874. ....... E.H.; Nepal
- ***palustre*** L., Spec. Plant. ed. 2: 1106. 1753. ....... E.H.; Nepal
- ***papillosum*** Lindb., Act. Soc. Sc. Fenn. 10: 280. 1872. ....... E.H.
- ***plumulosum*** Roell., Flora 69: 89. 1886. ....... E.H.; Nepal
- ***pseudocymbifolium*** C. Muell., Linnaea 38: 547. 1874. ....... E.H.; Nepal
- ***recurvum*** P. Beauv., Prodr. 88. 1805. ....... E.H.; Nepal
- ***squarrosum*** Crom. *in* Hopp., Bot. Zeit. 2: 324. 1803. .......W.H.; E.H.; Nepal
- ***subsecundum*** Nees *in* Sturm, Seutchl. Fl. 2(17): 3. 1819. ....... E.H.
- ***teres*** (Schimp.) Aongstr. *in* Hartm., Handb. Skand. Fl. ed. 8: 417. 1861. *Sphagnum squarrosum* var. *teres* Schimp., Vers. Entwickl. Torfm. 64. 1858. ....... E.H.

***Splachnobryum*** C. Muell., Verh. Zool. Bot. Ges. Wien 19: 503. 1869. (Splachnaceae)

L. Type : *S. obtusum* (Brid.) C. Muell.

- ***assamicum*** Dix., J. Bombay Nat. Hist. Soc. 39: 778. 1937. ....... E.H. Endemic
- ***bengalense*** Gang., Mosses E. India 4: 865. 1974. ....... G.P. Endemic
- ***flaccidum*** (Harv.) Braithw., Grevillea 1: 28. 1872. *Weisia flaccida* Harv. *in* Hook., Icon. Pl. Rar. 1: 18 *f.* 3. 1836. ....... G.P.; Nepal

– ***indicum*** Hamp. *et* C. Muell., Linnaea 37: 174. 1872.
....... W.H.; C.I.; G.P.; S.I.; Nepal

– ***procerrimum*** Dix. *et* P. Vard., Ann. Crypt. Exot. 1: 38. 2 *f.* 2. 1928.
....... W.H.; C.I.

– ***synoicum*** Robins., J. Hattori Bot. Lab. 27: 126-27. 1964.
....... W.H.; E.H.

***Stegonia*** Vent., Rev. Bryol. 10: 96. 1883. (Pottiaceae)

Type: *S. latifolia* (Schwaegr.) Vent.

– ***latifolia*** (Schwaegr.) Vent. *ex* Broth., Laubm. Fennosk. 145. 1923. *Weisia latifolia* Schwaegr. in Schultes, Reise Glockner 4. 1804.
....... W.H.

***Stenotheciopsis*** Fleisch., Musci Fl. Buitenzorg 4: 1477. 1923. (Hylocomiaceae)

– ***serrula*** (Mitt.) Fleisch., Musci Fl. Buitenzorg 4: 1477. 1923. *Stereodon serrula* Mitt., Musci Ind. Or. 114. 1859.
....... W.H.; E.H.

***Stereophyllum*** Mitt., Musci Ind. Or. 117. 1859. (Plagiotheciaceae)

Type: *S. indicum* (Bel.) Mitt.

– ***acuminatum*** Dix. *et* P. Vard., Ann. Crypt. Exot. 1: 46. 4 *f.* 4. 1928.
....... S.I.

– ***anceps*** (Bosch *et* Lac.) Broth., Nat. Pfl. 1(3): 898. 1907. *Hypnum anceps* Bosch *et* Lac., Bryol. Jav. 2: 161. 260. 1867.
....... E.H.; S.I.; C.I.

– ***confusum*** Ther., Annuair Cons. Jard. Bot. Geneve 20: 17. 1916.
....... S.I.; C.I.; Nepal

– ***decorum*** (Mitt.) Wijk *et* Marg., Taxon 9: 52. 1960. *Hypnum decorum* Mitt., Musci Ind. Or. 77. 1859. ....... W.H.; Nepal

– ***fulvum*** (Harv.) Jaeg., Ber. S. Gall. Naturw. Ges. 1877-78: 279. 1880. *Leskea fulva* Harv. *in* Hook., Icon. Pl. Rar. 1: 23 *f.* 3. 1836.
....... E.H.; S.I.; Nepal

– ***indicum*** (Bel.) Mitt., J. Linn. Soc. Bot. Suppl. 1: 117. 1859.
....... S.I.

- ***ligulatum*** Jaeg., Ber. S. Gall. Naturw. Ges. 1877-78: 277. 1880. ....... S.I.; C.I.
- ***setschwanicum*** Broth., Sitzungsber, Ak. Wiss. Wien Math. Nat. Kl. Abt. 1, 133: 581. 1924. ....... E.H.; Nepal
- ***subacuminatum*** Dix. *et* P. Vard., Ann. Crypt. Exot. 3: 182. 4. *f.* 2. 1930. ....... S.I.; C.I.
- ***tavoyense*** (Hook.) Jaeg., Ber. S. Gall. Naturw. Ges. 1877-78: 279. 1880. *Hypnum tavoyense* Hook., Icon. Pl. Rar. 1: 24 *f.* 1. 1836. ....... W.H.; S.I.; G.P.; Nepal
- ***wightii*** (Mitt.) Jaeg., Ber. S. Gall. Naturw. Ges. 1877-78: 279. 1880. *Hypnum wightii* Mitt., Musci Ind. Or. 82. 1859. ....... W.H.; E.H.; S.I.; C.I.; G.P.

***Struckia*** C. Muell., Arch. Ver. Freund. Naturg. Mecklenburg 47: 129. 1893. (Sematophyllaceae)

Type: *S. argentatum* (Mitt.) C. Muell.

- ***argentata*** (Mitt.) C. Muell., Arch. Ver. Freund. Naturg. Mecklenburg 47: 130. 1893. *Hypnum argentatum* Mitt., Musci Ind. Or. 77. 1859. ....... W.H.; E.H.; Nepal
- ***griffithi*** C. Muell., Arch. Ver. Freund. Naturg. Mecklenburg 47: 130. 1893. ....... E.H.

***Symblepharis*** Mont., Ann. Sc. Nat. Bot. ser. 2, 8: 252. 1837. (Dicranaceae)

Type: *S. helicophylla* Mont.

- ***reinwardtii*** (Doz. *et* Molk.) Lac., Bryol. Jav. 2: 225. 1870. *Dicranum reinwardtii* Doz. *et* Molk., Ann. Sc. Nat. Bot. ser. 3, 2: 303. 1844. ....... E.H.; Nepal
- ***vaginata*** (Hook.) Wijk *et* Marg., Taxon 8: 75. 1959. *Didymodon vaginatus* Hook., London J. Bot. 2: 5. 1840. ....... W.H.; E.H.; S.I.; Nepal

***Symphyodon*** Mont., Ann. Sc. Nat. Bot. ser. 2, 16: 279. 1941. (Symphyodontaceae)

Type: *S. perrottetii* Mont.

– ***asper*** (Mitt.) Jaeg., Ber., S. Gall. Naturw. Ges. 1876-77: 296. 1878. *Stereodon asper* Mitt., J. Linn. Soc. Bot. Suppl. 1: 110. 1859.
....... E.H.

– ***complanatus*** Dix., Rec. Bot. Surv. India 6: 65. 2. 1914.
....... E.H. Endemic

– ***echinatus*** (Mitt.) Jaeg., Ber. S. Gall. Naturw. Ges. 1876-77: 296. 1878. *Stereodon echinatus* Mitt., Musci Ind. Or. 110. 1859.
....... E.H.; Nepal

– ***erinaceus*** (Mitt.) Jaeg., Ber. S. Gall. Naturw. Ges. 1876-77: 296. 1878. *Stereodon erinaceus* Mitt., Musci Ind. Or. 110. 1859.
....... E.H.

– ***erraticus*** (Mitt.) Jaeg., Ber. S. Gall. Naturw. Ges. 1876-77: 296. 1878. *Stereodon erraticus* Mitt., Musci Ind. Or. 111. 1859.
....... E.H.; Nepal

– ***oblongifolius*** (Ren. *et* Card.) Broth., Nat. Pfl. 1(3): 877. 1907. *Leptohymenium oblongifolium* Ren. *et* Card., Bull. Soc. R. Bot. Belg. 34(2): 74. 1896. ....... E.H. Endemic

– ***orientalis*** (Mitt.) Broth. *ex* Par., Coll. 33. 1909. *Stereodon orientalis* Mitt., Musci Ind. Or. 111. 1859. ....... E.H.

– ***scabrisetus*** Dix., Rec. Bot. Surv. India 6: 66. 1914.
....... E.H. Endemic

***Symphysodontella*** Fleisch., Musci Fl. Buitenzorg 3: 688. 1908.
(Pterobryaceae)

– ***borii*** Dix., J. Bombay Nat. Hist. Soc. 39: 781. 1937.
....... E.H. Endemic

– ***involuta*** (Thwait. *et* Mitt.) Fleisch., Par. Coll. 33. 1909. *Pterobryum involutum* Thwait. *et* Mitt., J. Linn. Soc. Bot. 13: 315. 1873.
....... S.I.

– ***pilifolia*** Dix., J. Bombay Nat. Hist. Soc. 39: 782. 7 *f.* 8. 1937.
....... E.H. Endemic

– ***subulata*** Broth., Leafl. Philipp. Bot. 2(37): 654. 1909. ....... E.H.

– ***tortifolia*** Dix., J. Bombay Nat. Hist. Soc. 39: 782. 1937.
....... E.H. Endemic

***Syntrichia*** Brid., J. Bot. 1800, 1(2): 299. 1807. (Pottiaceae)

– ***princeps*** (De Not.) Mitt., Musci Ind. Or. 39. 1859. *Tortula princeps* De Not., Mem. R. Acc. Sc. Torino 40: 288. 1838. ....... W.H.; E.H.

***Syrrhopodon*** Schwaegr., Spec. Musc. Suppl. 2(1): 110. 1824. (Calymperaceae)

L. Type: *S. gardneri* (Hook.) Schwaegr.

– ***assamicus*** Robins., Bryologist 71: 84. 1968. ....... E.H. Endemic

– ***burmensis*** (Hamp.) Reese & Tan, Taxon 35(4): 693. 1986. *Calymperes burmense* Hamp. in Besch., Ann. Sci. Nat. bot. ser. 8, 1: 267, 279. 1896. ....... E.H.

– ***calymperoides*** Card. *et* P. Vard., Rev. Bryol. 49(3): 42. 9. 1922. ....... S.I.

– ***gardneri*** (Hook.) Schwaegr., Spec. Musc. Suppl. 2(1): 110. 1824. *Calymperes gardneri* Hook., Musci Exot. 2: 146. 1819. ....... W.H.; E.H.; S.I.; Nepal

– ***himalayanus*** Dix., Ann. Bryol. 3: 52. 1929. ....... W.H.

– ***larminatii*** Broth. *et* Par., Rev. Bryol. 28: 125. 1901. ....... E.H.

– ***leucophanoides*** Card. *et* P. Vard., Rev. Bryol. 49: 42. 10. 1922. ....... S.I.

– ***rufescens*** Hook. *et* Grev., Edinburgh J. Sc. 3: 227. 1826. ....... Nepal

– ***spiculosus*** Hook. *et* Grev., Edinburgh J. Sc. 3: 226. 1825. ....... W.H.

– ***strictus*** Thwait. *et* Mitt., J. Linn. Soc. Bot. 13: 299. 1873. ....... S.I.

– ***subconfertus*** Broth., Bot. Tidskrift 24: 117. 1901. ....... Andaman Is.

***Takakia*** Hatt. *et* Inoue, J. Hattori bot. Lab. 19: 133. 1958. (Takakiaceae)

Type: *T. lepidozioides* Hatt. *et* Inoue

– ***ceratophylla*** (Mitt.) Grolle, Osterr. Bot. Z. 110(4): 444. 1963. *Lepidozia ceratophylla* Mitt., J. Linn. Soc. Bot. 5: 103. 1861. ....... E.H.; Nepal

– ***lepidozioides*** Hatt. *et* Inoue, J. Hattori bot. Lab. 19: 133. 1958. ....... Nepal

***Taxiphyllum*** Fleisch., Musci F1. Buitenzong 4: 1434. 1923. (Hypnaceae)

– ***giraldii*** (C. Muell.) Fleisch., Musci Fl. Buitenzorg 4: 1435. 1923. *Plagiothecium giraldii* C. Muell., Nuov. Giorn. Bot. Ital. n. ser. 3: 114. 1896. ....... W.H.; Nepal

– ***maniae*** (Ren. *et* Par.) Fleisch., Musci Fl. Buitenzorg 4: 1436. 1923. *Isopterygium maniae* Ren.*et* Par., Rev. Bryol. 29: 84. 1902. ....... W.H.; Nicobar Is.; Nepal

– ***taxirameum*** (Mitt.) Fleisch., Musci Fl. Buitenzorg 4: 1435. 1923. *Stereodon taxirameus* Mitt., Musci Ind. Or. 105. 1859. ....... W.H.; E.H.; S.I.; C.I.; Nicobar Is.; Nepal

***Taxithelium*** Spruc. *ex* Mitt., J. Linn. Soc. Bot. 12: 21, 496. 1869. (Sematophyllaceae)

L. Type: *T. planum* (Brid.) Mitt.

– ***celatum*** (Mitt.) Broth., Nat. Pfl. 1(3): 1237. 1909. *Stereodon celatus* Mitt., J. Linn. Soc. Bot. Suppl. 1: 99. 1859. ....... Nepal

– ***instratum*** (Brid.) Broth. *in* Ren. *et* Card., Rev. Bryol. 28: 110. 1901. *Hypnum instratum* Brid., Bryol. Univ. 2: 391. 1827. ....... S.I.

– ***kerianum*** (Broth.) Broth., Nat. Pfl. 1(3): 1091. 1908. *Trichosteleum kerianum* Broth., Oefv. Finsk. Vet. Soc. Foerh. 33: 108. 1891. ....... E.H.; Nicobar Is.

– ***laeviusculum*** Dix., J. Bombay Nat. Hist. Soc. 39: 793. 1937. ....... E.H. Endemic

– ***nepalense*** (Schwaegr.) Broth., Monsunia 1: 51. 1899. *Hypnum nepalense* Schwaegr., Spec. Musc. Suppl. 3(1): 226. 1828. ....... E.H.; S.I.; C.I.; Andaman & Nicobar Is.; Nepal

– ***vernieri*** (Dub.) Besch., Bull. Soc. Bot. France 45: 123. 1898. *Hypnum vernieri* Dub., Flora 58: 285. 1875. ....... Andaman & Nicobar Is.

***Tayloria*** Hook., J. Sc. Arts 2(3): 144. 3. 1816. (X). (Splachnaceae)

Type: *T. splachnoides* (Schwaegr.) Hook.

– ***froelichiana*** (Hedw.) Mitt. *in* Broth., Nat. Pfl. 1(3): 502. 1903. *Splachnum froelichianum* Hedw., Spec. Musc. 52. 1801.
....... W.H.

– ***hornschuchii*** (Gerv. *et* Arnott) Broth., Nat. Pfl. 1(3): 502. 1903. *Dissodon hornschuchii* Grev. *et* Arnott, Mem. Wernerian Nat. Hist. Soc. 5(2): 468. 13 *f.* 34-38. 1826. ....... Nepal

– ***imbricata*** Thwait. *et* Mitt., J. Linn. Soc. Bot. 13: 305. 1873
....... S.I.

– ***indica*** Mitt., J. Linn. Soc. Bot. Suppl. 1: 57. 1859.
....... W.H.; E.H.; Nepal

– ***jacquemontii*** (B.S.G.) Mitt., J. Linn. Soc. Bot. Suppl. 1: 57. 1859. *Dissodon jacquemontii* B.S.G., Bryol. Eur. 3: 192. 1844.
....... W.H.

– ***subglabra*** (Griff.) Mitt., J. Linn. Soc. Bot. Suppl. 1: 57. 1859. *Orthodon subglaber* Griff., Calcutta J. Nat. Hist. 2: 483. 1842.
....... E.H.; S.I.; Nepal

– ***tenella*** Mitt., J. Linn. Soc. Bot. Suppl. 1: 57. 1859. ....... W.H.

***Tetraplodon*** B.S.G., Bryol. Eur. 3: 211. 1844. (Splachnaceae)

– ***angustatus*** (Hedw.) B.S.G., Bryol. Eur. 3: 214. 288. 1844. *Splachnum angustatum* L.*f.* *ex* Hedw., Spec. Musc. 51. 1801.
....... E.H.

***Thamnobryum*** Nieuwl., Am. Midland Natural. 5: 50. 1917.
(Neckeraceae)

– ***alopecurum*** (Hedw.) Nieuwl., Am. Midland Natural. 5: 50. 1917. *Hypnum alopecurum* L. *ex* Hedw., Spec. Musc. 267. 1801.
....... W.H.; S.I.; Nepal

– ***fruticosum*** (Mitt.) Gang., Mosses E India 5: 1447. 1976. *Neckera fruticosa* Mitt., J. Linn. Soc. Bot. Suppl. 1: 122. 1859.
....... E.H.; Nepal

– ***latifolium*** (Bosch *et* Lac.) Nieuwl., Am. Midland Natural. 5: 50. 1917. *Porotrichum latifolium* Bosch *et* Lac., Bryol. Jav. 2: 69. 188. 1863. ....... W.H.; Nepal

– ***macrocarpum*** (Brid.) Gang., Mosses E. India 5: 1454. 1976. *Neckera macrocarpa* Brid., Bryol. Univ. 2: 233. 1827. ....... Nepal

– ***parvulum*** (Mitt.) R.S. Chopra, Taxon. Indian Mosses: 377. 1975. *Neckera parvula* Mitt., J. Proc. Linn. Soc. Bot. Suppl. 1: 123. 1859. ....... S.I.

– ***schmidii*** (C. Muell.) R. S. Chopra, Taxon. Indian Mosses: 377. 1975. *Hypnum schmidii* C. Muell., Bot. Zeit. 12: 574. 1854. ....... S.I.

– ***subseriatum*** (Hook.) R.S. Chopra, Taxon. Indian Mosses : 377. 1975. *Neckera subseriata* Hook., Icon. Pl. 1: pl. 21 f. 7. 1836. ....... W.H.; E.H.; S.I.; Nepal

***Theriotia*** Card., Beih. Bot. Centralbl. 17: 8. 1904. (Buxbaumiaceae)

Type: *T. lorifolia* Card.

– ***kashmirensis*** Robins., Bryologist 68: 314. 1965. ....... W.H.

– ***lorifolia*** Card., Beih. Bot. Centralbl. 17: 8. 5. 1904. ....... W.H.

***Thiemea*** C. Muell., Bot. Centralbl. 7: 346. 1881. (Leucobryaceae)

Type: *T. hampeana* C. Muell.

– ***hampeana*** C. Muell., Bot. Centralbl. 7: 346. 1881. ....... S.I.; Nepal

***Thuidium*** B.S.G., Bryol. Eur. 5: 157. 1852. (Thuidiaceae)

Type: *T. tamariscinum* (Hedw.) B.S.G.

– ***assimile*** (Mitt.) Jaeg., Ber. S. Gall. Naturw. Ges. 1876-77: 260. 1878. *Leskea assimilis* Mitt., J. Linn. Soc. Bot. Suppl. 1: 133. 1859. ....... W.H.; E.H.

– ***brotheri*** Salm., J. Bot. 39: 153. 421. 1901. ....... E.H.; C.I.

– ***contortulum*** (Mitt.) Jaeg., Ber. S. Gall. Naturw. Ges. 1876-77: 256. 1878. *Leskea contortula* Mitt., J. Linn. Soc. Bot. Suppl. 1: 134. 1859. ....... W.H.; E.H.

– ***cymbifolium*** (Doz. *et* Molk.) Doz. *et* Molk., Bryol. Jav. 2: 115. 221. 1865. *Hypnum cymbifolium* Doz. et Molk., Ann. Sc. Nat. Bot. ser. 3, 2: 306. 1844. ....... W.H.; E.H.; S.I.; C.I.; Nicobar Is.; Nepal

var. ***brachymenium*** (Herz.) Vohra, Rec. Bot. Surv. India 23: 122. 1983. *T. brachymenium* Herz., Hedwigia 65. 164. 1925. ....... Nepal

– ***delicatulum*** (Hedw.) Mitt., J. Linn. Soc. Bot. 12: 578. 1869. *Hypnum delicatulum* Hedw., Spec. Musc. 260. 1801. ....... Nepal

– ***glaucinum*** (Mitt.) Bosch *et* Lac., Bryol. Jav. 2: 117. 222. 1865. *Leskea glaucina* Mitt., J. Linn. Soc. Bot. Suppl. 1: 133. 1859. ....... W.H.; E.H.; S.I.; Nepal

– ***haplohymenium*** (Harv.) Jaeg., Ber. S. Gall. Naturw. Ges. 1876-77: 251. 1878. *Hypnum haplohymenium* Harv., London J. Bot. 2: 21. 1843.

var. ***haplohymenium*** ....... W.H.; E.H.; Nepal

– ***investe*** (Mitt.) Jaeg., Ber. S. Gall. Naturw. Ges. 1876-77 : 252. 1878. *Hypnum investe* Mitt., Kew J. Bot. 8: 355. 1856. ....... E.H.

– ***kiasense*** Williams, Bull. N. Y. Bot. Gard. 8: 363. 1914. ....... E.H.

– ***koelzii*** Robins., Bryologist 71: 92. 1968. ....... W.H.; E.H.; C.I.; Nepal

– ***meyenianum*** (Hamp.) Doz. *et* Molk., Bryol. Jav. 2: 121. 224. 1865. *Hypnum meyenianum* Hamp., Icon. Musc. 8. 1844. ....... W.H.; E.H.; S.I.; Nicobar Is.; Nepal

– ***minusculum*** (Mitt.) Jaeg., Ber. S. Gall. Naturw. Ges. 1876-77: 256. 1878. *Leskea minuscula* Mitt., J. Linn. Soc. Suppl. 1: 134. 1859. ....... W.H.; E.H.; Nepal

– ***orientale*** Mitt. *ex* Dix., J. Bot. 51: 329. 1913. ....... W.H.; E.H.

– ***philibertii*** Limpr., Laubm. Deutschl. 2: 835. 1895. ....... W.H.; Nepal

– ***recognitum*** (Hedw.) Lindb., Not. Saellsk. F. Fl. Fenn. Foerh. 13: 416. 1874. *Hypnum recognitum* Hedw., Spec. Musc. 261. 1801. ....... Nepal

– ***sparsifolium*** (Mitt.) Jaeg., Ber. S. Gall. Naturw. Ges. 1876-77: 248. 1878. *Leskea sparsifolia* Mitt., J. Linn. Soc. Bot. Suppl. 1: 135. 1859. ....... W.H.; E.H.

– ***squarrosulum*** Ren. *et* Card., Bull. Soc. R. Bot. Belg. 38(1): 31. 1900.

var. ***squarrosulum*** ....... W.H.; E.H.; Nepal

– ***stevensii*** Ren. *et* Card., Bull. Soc. R. Bot. Belg. 38(1): 33. 1900. ....... E.H. Endemic

– ***subpellucens*** Dix., J. Bombay Nat. Hist. Soc. 39: 787. 1937.
....... E.H. Endemic

– ***talongense*** Besch., Ann. Sc. Nat. Bot. ser. 7, 15: 81. 1892.
....... I.O.

– ***tamariscellum*** (C. Muell.) Bosch et Lac., Bryol. Jav. 2: 20. 1865. *Hypnum tamariscellum* C. Muell., Bot. Zeit. 12: 573. 1854.
.......W.H.; E.H.; S.I.; Nepal

– ***venustulum*** Besch., Ann. Sc. Nat. Bot. sec. 7, 15: 78. 1892.
....... E.H.

– ***vestitissimum*** Besch., Ann. Sc. Nat. Bot. ser. 7, 15: 79. 1892.
....... W.H.; Nepal

***Timmia*** Hedw., Spec. Musc. 176. 1801. (Timmiaceae)

Type: *T. megapolitana* Hedw.

– ***austriaca*** Hedw., Spec. Musc. 176. 42 f. 1-7. 1801 ....... W.H.

– ***bavarica*** Hessl., Musc. Fr. Gen. 19. 3. 1822. ....... W.H.

– ***megapolitana*** Hedw., Spec. Musc. 176. 1801. ....... W.H.

***Timmiella*** (De Not.) Limpr., Laubm. Deutschl. 1: 590. 1888.
(Pottiaceae)

Type: *T. anomala* (B.S.G.) Limpr.

– ***anomala*** (B.S.G.) Limpr., Laubm. Deutschl. 1: 592. 173. 1888. *Barbula anomala* B.S.G., Bryol. Eur. 2: 107, 169. 1842.
....... W.H.; S.I.

– ***barbuloides*** (Brid.) Moenk., Laubm. Eur. 273. 1927. *Trichostomum barbuloides* Brid., Spec. Musc. 1: 233. 1806. ....... W.H.

– ***diminuta*** (C. Muell.) Chen, Hedwigia 80: 176. 1941. *Trichostomum diminutum* C. Muell., Nuov., Giorn. Bot. Ital. n. ser. 5: 177. 1898.
....... W.H.

***Tortella*** (Lindb.) Limpr., Laubm. Deutschl. 1: 599. 1888. (Pottiaceae)

L. Type: *T. caespitosa* (Schwaegr.) Limpr.

– ***alpicola*** Dix., Ann. Bryol. 3: 54. 1929. ....... W.H.

– ***ceylonensis*** Fleisch. *ex* Dix., Not. R. Bot. Gard. Edinburgh 19: 284. 5. 1938. ....... I.O.

– ***fragilis*** (Hook. *et* Wils.) Limpr., Laubm. Deutschl. 1: 606. 1888. *Didymodon fragilis* Hook. *et* Wils. *in* Drumm., Musci Bor. Am. 127. 1828. ....... W.H.; E.H.

– ***tortuosa*** (Hedw.) Limpr., Laubm. Deutschl. 1: 604. 1888. *Tortula tortuosa* Hedw., Spec. Musc. 124. 1801. ....... W.H.; Nepal

– ***undulatifolia*** Dix., 150th Ann. Bot. Gard. Calcutta 180. 1942. ....... I.O.

– ***walkeri*** (Broth.) Zand., Bull. Buffalo Soc. Nat. Sc. N. Y. 32: 104. 1993. *Hyophila walkeri* Broth., Rec. Bot. Surv. India 1: 317. 1899. ....... S.I.; C.I. Endemic

***Tortula*** Hedw., Spec. Musc. 122. 1801. (Pottiaceae)

L. Type: *T. subulata* Hedw.

– ***brandisii*** (C. Muell.) Broth., Nat. Pfl. 1(3): 432. 1902. *Barbula brandisii* C. Muell., Flora 61: 82. 1878. ....... W.H.

– ***caninervis*** (Mitt.) Broth., Nat. Pfl. 1(3): 435. 1902. *Syntrichia caninervis* Mitt., J. Linn. Soc. Bot. Suppl. 1: 39. 1859. ....... I.O.

– ***inermis*** (Brid.) Mont., Arch. Bot. 1: 136. 4 f. 4. 1832. *Syntrichia subulata* (Hedw.) Culm. in Amann var. *inermis* Brid., Bryol. Univ. 581. 1826. ....... W.H.

– ***microphylligera*** Broth. *in* Bruehl, Rec. Bot. Surv. India 13(1): 127. 1931. ....... E.H.

– ***montana*** (Nees) Lindb., Musc. Scand. 20. 1879. *Syntrichia montana* Nees in Raab, Flora 2: 301. 1819. ....... W.H.

– ***mucronifolia*** Schwaegr., Spec. Musc. Suppl. 1(1): 136. 34. 1811. ....... I.O.

– ***muralis*** Hedw., Spec. Musc. 123. 1801. ....... W.H.; E.H.

– ***nigrescens*** Griff. *in* Gang., Mosses E. India 3: 700. 1972. ....... I.O.

– ***princeps*** De Not., Mem. R. Acc. Sc. Torino 40: 288. 1838. ....... W.H.; E.H.

– ***pseudo-princeps*** Dix., Ann. Bryol. 3: 58. 1930. ....... W.H.

– ***rubripila*** Dix., Ann. Bryol. 3: 58. 1930. ....... W.H.

– ***ruralis*** (Hedw.) Gaertn., Meyer *et* Scherb., Oek Techn. Fl. Wetterau 3(2): 91. 1802. *Barbula ruralis* Hedw., Spec. Musc. 121. 1801. ....... W.H.

– ***schmidii*** (C. Muell.) Broth., Nat. Pfl. 1(3): 434. 1902. *Barbula schmidii* C. Muell., Bot. Zeit. 11: 58. 1853. ....... S.I.

– ***subulata*** Hedw., Spec. Musc. 122. 27 *f.* 1-3. 1801. ....... W.H.

– ***websteri*** Robins., Bryologist 68: 316. 1965. ....... E.H.

***Trachycarpidium*** Broth., Nat. Pfl. 1(3): 383. 1901. (Pottiaceae)

Type: *T. verrucosum* (Besch.) Broth.

– ***tisserantii*** Dix. *et* P. Vard., Arch. Bot. 1(3): 44, 13. 1928. ....... S.I.

***Trachyloma*** Brid., Bryol. Univ. 2: 277. 1827. (Pterobryaceae)

Type: *T. planifolium* (Hedw.) Brid.

– ***indicum*** Mitt., J. Linn. Soc. Bot. Suppl. 1: 91. 1859. ....... E.H.; S.I.

***Trachyphyllum*** Gepp *in* Hiern, Cat. Welw. Afr. Pl. 2(2): 298. 1901. (Thuidiaceae)

– ***elongatum*** Dix. *et* P. Vard., Rev. Bryol. 52: 40. 1925. ....... I.O.

– ***fragilifolium*** Dix., Not. R. Bot. Gard. Edinburgh 19(95): 297. 1938. ....... I.O.

– ***inflexum*** (Harv,) Gepp *in* Hiern, Cat. Welw. Afr. Pl. 2(2): 299. 1901. *Hypnum inflexum* Harv. *in* Hook., Icon. Pl. Rar. 1 : t. 24(6). 1836. ....... E.H.; S.I.; C.I.; Nepal

– ***jeyporense*** Thér. *et* Dix. in P. Vard., Ann. Crypt. Exot. 1: 46. 4 f. 3. 1928. ....... C.I. Endemic

***Trachypodopsis*** Fleisch., Hedwigia 45: 64. 1905. (Trachypodaceae)

– ***auriculata*** (Mitt.) Fleisch., Hedwigia 45: 67. 1906. *Trachypus auriculatus* Mitt., J. Linn. Soc. Bot. Suppl. 1: 129. 1859. ....... W.H.; E.H.; S.I.; Nepal

– ***serrulata*** (P. Beauv.) Fleisch., Hedwigia 45: 67. 1906. *Pilotrichum serrulatum* P. Beauv., Prodr. 83. 1805. ....... W.H.; E.H.; S.I.; Andaman Is.; Nepal

***Trachypus*** Reinw. *et* Hornsch., Nov. Act. Ac. Leop. Car. 14(2): 708. 1829. (Trachypodaceae)

Type: *T. bicolor* Reinw. *et* Hornsch.

- ***appendiculatus*** (Ren. *et* Card.) Broth., Nat. Pfl. 1(3): 829. 1906. *Papillaria appendiculata* Ren. et Card., Bull. Soc. R. Bot. Belg. 32(1): 108. 1894. ....... Himalayas
- ***bicolor*** Reinw. *et* Hornsch., Nov. Act. Ac. Leop. Car. 14(2): 39. 1829. ....... W.H.; E.H.; S.I.; Nepal
- ***humilis*** Lindb., Act. Soc. Sc. Fenn. 10: 230. 1872. ....... S.I.

***Trematodon*** Michx., Fl. Bor. Am. 2: 289. 1803. (Dicranaceae)

Type : *T. longicollis* Michx.

- ***ambiguus*** (Hedw.) Hornsch., Flora 2: 88. 1819. *Dicranum ambiguum* Hedw., Spec. Musc. 150. 1801. ....... Nepal
- ***assamensis*** Broth. *in* Roth, Aussereur. Laubm. 278, 292. 28 *f.*1. 1911. ....... E.H. Endemic
- ***brevicalyx*** Dix. *in* Mathur, J. Indian Bot. Soc. 4: 287-97. 1925. ....... G.P.
- ***capillifolius*** C. Muell. *ex* Roth, Aussereur. Laubm. 296. 28 *f.* 3. 1911. ....... W.H.
- ***ceylonensis*** C. Muell., Bot. Zeit. 22 : 350. 1864. ....... E.H.; S.I.; C.I
- ***conformis*** Mitt., J. Linn. Soc. Bot. Suppl. 1: 12. 1859. ....... W.H.; E.H.; S.I.; Nepal
- ***hookeri*** C. Muell., Bot. Zeit. 22: 350. 1864. ....... E.H.; Nepal
- ***kurzii*** Hamp. *ex* Gang., Bull. Bot. Soc. Bengal 14: 12. 30. 1960. ....... E.H.
- ***longicollis*** Michx., Fl. Bor. Am. 2: 289. 1803. ....... W.H.; E.H.; S.I.; G.P.
- ***megapophysatus*** C. Muell., Bot. Zeit. 22: 350. 1864. ....... E.H. Endemic
- ***schmidii*** C. Muell., Bot. Zeit. 11: 40. 1853. ....... S.I.
- ***subulosus*** Griff., Calcutta J. Nat. Hist. 2: 493. 1842. ....... W.H.; E.H.; P.

***Trichosteleum*** Mitt., J. Linn. Soc. Bot. 10: 181. 1868. (Sematophyllaceae)

L. Type: *T. fissum* Mitt.

– ***boschii*** (Doz. *et* Molk.) Jaeg., Ber. S. Gall. Natturw. Ges. 1876-77: 421. 1878. *Hypnum boschii* Doz. *et* Molk., Ann. Sc. Nat. Bot. ser. 3. 2: 306. 1844. ....... E.H.; Nicobar Is.; Nepal

– ***glauco-virens*** (Mitt.) Broth., Nat. Pfl. 1(3): 1119. 1908. *Stereodon glauco-virens* Mitt., Musci Ind. Or. 103. 1859. ....... E.H. Endemic

– ***hamatum*** (Doz. *et* Molk.) Jaeg., Ber. S. Gall. Naturw. Ges. 1876-77: 420. 1878. *Hypnum hamatum* Doz. *et* Molk., Ann. Sc. Nat. Bot. ser. 3. 2: 307. 1844. ....... E.H.; S.I.

– ***luxurians*** (Doz. *et* Molk.) Broth., Nat. Pfl. 1(3): 1119. 1908. *Hypnum luxurians* Doz. *et* Molk., Ann. Sc. Nat. Bot. ser. 3. 2: 307. 1844. ....... E.H.; Nicobar Is.

– ***monostictum*** (Thwait. *et* Mitt.) Broth., Rec. Bot. Surv. India 1: 326. 1898. *Sematophyllum monostictum* Thwait. et Mitt., J. Linn. Soc. Bot. 13: 318. 1873. ....... E.H.; S.I.; Nicobar Is.

– ***punctipapillosum*** Par. *ex* Gang., Mosses E. India. 8: 1913. 1980. ....... Andaman Is. Endemic

– ***stereodontoides*** Broth. *ex* Gang., Mosses E. India 8: 1915. 1980. ....... E.H.

– ***stissophyllum*** (Hamp.) Jaeg., Ber. S. Gall. Naturw. Ges. 1876-77: 417. 1878. *Hypnum stissophyllum* Hamp., Syn. 2: 273. 1851. ....... E.H.

***Trichostomum*** Bruch, Flora 12: 396. 1829. (Pottiaceae)

L. Type: *T. brachydontium* Bruch

– ***bombayense*** C. Muell., Syn. Musc.Frond. 2: 628. 1851. ....... C.I.

– ***brachydontium*** Bruch *in* F. A. Muell., Flora 12: 393. 3. 1829. ....... W.H.; E.H.

– ***hyalinoblastum*** (Broth.) Broth., Nat. Pfl. 1(3): 394. 1902. *Tortella hyalinoblasta* Broth., Rec. Bot. Surv. India 1: 318. 1899. ....... S.I.

– ***lillei*** Dix., Rec. Bot. Surv. India 9(5): 306, 1 *f.* 4. 1926. .......W.H.

– ***minusculum*** Dix. *et* P. Vard., Arch. Bot. 1 (8-9): 165. 8 f. 7. 1927. ...... S.I.

– ***orthodontum*** (C. Muell.) Broth., Nat. Pfl. 1(3): 394. 1902. *Barbula orthodonta* C. Muell., Bot. Zeit. 11: 57. 1853. ....... S.I.

– ***perannulatum*** Dix. *et* P. Vard., Ann. Crypt. Exot. 1: 280. 2. 1928. ....... S.I.

– ***subminusculum*** Dix. *et* P. Vard., Rev. Bryol. Lich. 15: 145. 1946. ....... S.I.

– ***uncifolium*** Dix., 150th Ann. Bot. Gard. Calcutta 180. 1942. ....... W.H.

***Trigonodictyon*** Dix. et P. Vard., Ann. Crypt. Exot. 1: 40. 1928. (Orthotrichaceae)

Type: *T. indicum* Dix. *et* P. Vard.

– ***indicum*** Dix. et P. Vard., Ann. Crypt. Exot. 1: 40. 1928. ....... S.I.

***Trismegistia*** (C. Muell.) C. Muell., Flora 82: 474. 1896 (x). (Sematophyllaceae)

Type: *T. subauriculata* C. Muell.

– ***lancifolia*** (Harv.) Broth., Nat. Pfl. 1(3): 1078. 1908. *Neckera lancifolia* Harv. *in* Hook., Icon. Pl. Rar. 1: 21. 1836. ....... Nepal

***Trolliella*** Herz., Ann. Bryol. 12: 94. 1939. (Sematophyllaceae)

Type: *T. euendostoma* Herz.

– ***euendostoma*** Herz., Ann. Bryol. 12: 94. 13. 1939. ....... E.H. Endemic

***Tuerckheimia*** Broth., Oefv. F. Vet. Soc. Foerh. 52A(7): 2. 1970. (Pottiaceae)

Type: *T. guatemalensis* Broth.

– ***svihlae*** (Bartr.) Zand., Bull. Buffalo Soc. Nat. Sci. NY 32: 94. 1993. *Trichostomum svihlae* Bartr., Rev. Bryol. Lichenol. 23: 245. 1954. ....... E.H.

***Ulota*** Mohr, Ann. Bot. 2 : 540. 1806. (Orthotrichaceae)

L. Type : *U. crispa* (Hedw.) Brid.

– ***robusta*** Mitt., J. Linn. Soc. Bot. Suppl. 1: 49. 1859. ....... E.H.; Nepal

– ***schmidii*** (C. Muell.) Mitt., J. Linn. Soc. Bot. Suppl. 1: 49. 1859. *Orthotrichum schmidii* C. Muell., Bot. Zeit. 11.60. 1853. ....... S.I.

***Vesicularia*** (C. Muell.) C. Muell., Bot. Jahrb. 23: 330. 1896. (Hypnaceae)

L. Type: *V. meyeniana* (Hamp.) Broth.

– ***firma*** Dix. *et* P. Vard., Ann. Crypt. Exot. 3: 184. 4 *f.* 7.1930. ....... S.I.; C.I.

– ***kurzii*** (Lac.) Broth., Nat. Pfl. 1(3): 1095. 1908. *Hypnum kurzii* Lac., Bryol. Jav. 2: 183. 281. 1867. ....... S.I.; C.I.

– ***levieri*** Card. *in* Dix., J.Indian Bot. 2: 187. 10: 1921. ....... C.I.; Andaman Is.

– ***montagnei*** (Bel.) Broth., Nat. Pfl. 1(3): 1094. 1908. *Pterygophyllum montagnei* Bel., Voyag. Ind. Or. Bot. 2 (Crypt.): 85. 1834. ....... W.H.; E.H.; G.P.

– ***nitidula*** Card. *et* P. Vard., Rev. Bryol. 51: 10. 15. 1924. ....... S.I.

– ***perreticulata*** Broth. *ex* Dix., J. Bot. 53: 295. 1915. ....... S.I.

– ***reticulata*** (Doz. *et* Molk.) Broth., Nat. Pfl. 1(3): 1094. 1908. *Hypnum reticulatum* Doz. *et* Molk., Ann. Sc. Nat. Bot. ser. 3. 2: 309. 1844. ....... W.H.; E.H.; S.I.; Nepal

– ***subpilicuspis*** Card. *et* P. Vard., Rev. Bryol. 50(3): 79. 1923. ....... S.I.

– ***subscaturiginosa*** Fleisch., Musci Fl. Buitenzorg 4: 1454. 1923. ....... W.H.

– ***succosa*** (Mitt.) Broth., Nat. Pfl. 1(3): 1094. 1908. *Stereodon succosus* Mitt., Musci Ind. Or. 101. 1859. ....... W.H.; E.H.; C.I.; Nepal

***Voitia*** Hornsch., Voitia Systylio Nov. Musc. Fr. Gen. 5, 9. 1818. (Splachnaceae)

Type: *V. nivalis* Hornsch.

– ***hookeri*** Mitt., J. Linn. Soc. Bot. Suppl. 1: 56. 1859. ....... E.H.

***Warburgiella*** C. Muell. in Broth., Monsunia 1: 176. 1900. (Sematophyllaceae)

Type: *W. cupressinoides* Broth.

– ***isopterygioides*** Dix. *et* P. Vard., Ann. Crypt. Exot. 3 : 183. 6. 1930. ....... S.I.

– ***leptocarpa*** (Schwaegr.) Fleisch., Musci F1. Buitenzorg 4: 1253. 1923. *Hypnum leptocarpon* Schwaegr., Spec. Musc. Suppl. 2(2): 171. 200. 1827. ....... S.I.

– ***leptorhynchoides*** (Mitt.) Fleisch., Musci F1. Buitenzorg 4: 1253. 1923. *Stereodon leptorhynchoides* Mitt., J. Linn. Soc. Bot. Suppl. 1: 103. 1859. ....... S.I.

– ***perviridis*** Dix. *et* P. Vard., Arch. Bot. 1(8-9): 176 *f.* 5. 1927. ....... S.I.

***Weissia*** Hedw., Spec. Musc. 168. 1801. (Pottiaceae)

L. Type : *W. controversa* Hedw.

– ***controversa*** Hedw., Spec. Musc. 67. 1801. ....... W.H.; S.I.

– ***ghatensis*** Dix. *et* P. Vard., Ann. Crypt. Exot. 3: 174. 2 *f.* 4. 1930. ....... S.I.; C.I.

– ***macrospora*** Card. *et* P. Vard., Rev. Bryol. 49: 43. 12. 1922. ....... S.I.

– ***rutilans*** (Hedw.) Lindb., Oefv. K. Vet. Ak. Foerh. 20 : 417. 1863. *Gymnostomum rutilans* Hedw., Spec. Musc. 37. 3f. 8-11. 1801. ....... W.H.

– ***wimmeriana*** (Sendtn.) B.S.G., Bryol. Eur. 1: 64. 20. 1846. *Gymnostomum wimmerianum* Sendtn., Flora 23: 59. 1840. ....... W.H.

***Wijkia*** Crum, Bryologist 74: 170. 1971. (Sematophyllaceae)

– ***baculifera*** (Dix.) Crum, Bryologist 74: 171. 1971. *Acanthocladium baculiferum* Dix., J. Bombay Nat. Hist. Soc. 39: 791. 1937. ....... E.H.

– ***deflexifolia*** (Ren. *et* Card.) Crum, Bryologist 74: 171. 1971. *Acanthocladium deflexifolium* Ren. *et* Card., Bull. Soc. R. Bot. Belg. 41(1): 92. 1905. ....... E.H.

– ***laxitexta*** (Ren. *et* Card.) Crum, Bryologist 74: 172. 1971. *Rhaphidostegium laxitextum* Ren. *et* Card., Bull. Soc. R. Bot. Belg. 34(2): 76. 1896. ....... E.H. Endemic

– ***lepida*** (Mitt.) Crum, Bryologist 74: 172. 1971. *Stereodon lepidus* Mitt., Musci Ind. Or. 98. 1859. ....... E.H.; Nepal

– ***penicillata*** (Mitt.) Crum, Bryolotist 74: 173. 1971. *Stereodon penicillatus* Mitt., Musci Ind. Or. 112. 1859. ....... E.H.; Nepal

– ***sikkimense*** (Broth.) R.S. Chopra, Taxon. Indian Mosses 502. 1975. ....... E.H.

– ***surcularis*** (Mitt.) Crum, Bryologist 74: 173. 1971. *Stereodon surcularis* Mitt., Musci Ind. Or. 112. 1859. ....... E.H.; Nepal

– ***tanytricha*** (Mont.) Crum, Bryologist 74: 174. 1971. *Hypnum tanytrichum* Mont., Ann. Sc. Nat. Bot. ser. 3, 4: 88. 1845. ....... W.H.; E.H.; C.I.

***Wilsoniella*** C. Muell., Bot. Centralbl. 7: 345. 1881. (Dicranaceae)

– ***decipiens*** (Mitt.) Alston *in* Dix., J. Bot. 68: 2. 1930. *Trematodon decipiens* Mitt., Musci Ind. Or. 13. 1859. ....... S.I.; C.I.; Andaman & Nicobar Is.

***Zygodon*** Hook. *et* Tayl., Musc. Brit. 70. 1818. (Orthotrichaceae)

Type: *Z. conoideus* (Dicks.) Hook. *et* Tayl.

– ***acutifolius*** C. Muell., Bot. Zeit. 11: 58. 1853. ....... S.I.

– ***brevisetus*** Wils. *ex* Mitt., J. Linn. Soc. Bot. Suppl. 1: 47. 1859. ....... E.H.

– ***erosus*** Mitt., J. Linn. Soc. Bot. 22: 305. 1886. ....... S.I.

– ***humilis*** Thwait. *et* Mitt., J. Linn. Soc. Bot. 13: 304. 1873. ....... S.I.

– ***intermedius*** B.S.G., Bryol. Eur. 3: 41. 1838. ....... S.I.

– ***obtusifolius*** Hook., Musci Exot. 2: 159. 1819. ....... E.H.; S.I.; Nepal

– ***reinwardtii*** (Hornsch.) Braun *in* B.S.G., Bryol. Eur. 3: 41. 1838. *Syrrhopodon reinwardtii* Hornsch., Nov. Act. Ac. Leop. Car. 14(2): 700. 39. 1829. ....... S.I.

– ***tetragonostomus*** Braun *ex* B.S.G., Bryol. Eur. 3: 42. 1838. ....... S.I.

– ***viridissimus*** (Dicks.) R. Br, Trans. Linn. Soc. 12: 575. 1819. *Bryum viridissimum* Dicks., Fl. Crypt. Brit. F. 4: 9. 1801. ....... E.H.; Nepal

– ***wightii*** Dix. *et* Malta, Not. R. Bot. Gard. Edinburgh 19: 288, 8. 1938. ....... S.I.

# REFERENCES


Aziz, M.N. & J.N. Vohra 1983. A new species of *Anoectangium* Schwaegr. (Musci) from India. - *Bull. Bot. Surv.* India **25**(1-4): 239-240.

Aziz, M.N. & J.N. Vohra 1988 (1990). Note on the identity of *Hymenostylium xanthocarpum* (Hook.) Brid. - *Bull. Bot. Surv. India.* **30**: 185-187.

Bapna, K.R. & B. L. Choudhary 1989. Mosses of Rajasthan II. Dicranales and Pottiales. - *J. Indian Bot. Soc.* **68**: 379-388.

Bartram, E.B. 1955. North-western Himalayan mosses. I. *Bull. Torr. Bot. Club,* **82**(1): 22-29.

Brotherus, V. F. 1898. Contributions to the Bryological Flora of the North-western Himalayas. - *Acta Soc. Sci. Fenn.* **24**(2): 1-46.

Brotherus, V.F. 1898. Contributions to the Bryological Flora of Southern India. Report on a collection of Mosses made by Dr. J.L. Walker in Coorg. - *Rec. Bot. Surv. India* **1**(12): 311-329.

Brotherus, V.F. 1901-1909. Musci in *Die Naturlichen Pflanzen familien* by Engler & Prantl. **1**(3) : 277- 1246.

Brotherus, V.F. 1924-1925. Musci (Laubmoose), In: A. Engler and Prantl (eds.), *Die Naturlichen Pflanzenfamilien* 11. Leipzig.

Brotherus, V.F. 1928. Contributions a la flore bryologique de Cachmire. - *Annls bryol.* **1**: 24-48.

Brotherus, V.F. & I. Theriot 1924. Musci Madurensis IV. *Revue bryol.* **51**: 10-14.

Bruehl, P. 1931. A Census of Indian Mosses. - *Rec. Bot. Surv. India* **13**(1): 1-135.

Bruehl, P. & N. Sarkar 1929. The Mosses of Bengal. Part 1- *J. Dep. Sci. Calcutta Univ.* **10**: 1-12.

Buck, W.R. & D.H. Vitt. 1986. Suggestions for a new familial classification of pleurocarpous mosses. - *Taxon* **35**: 21-60.

Chavan, A.R., S. Mehta & A.R. Mehta 1961. A preliminary report of mosses from Pavagadh in Gujrat. - *J. Univ. Baroda* **10** : 35-42.

Chopra, N. 1961. Mosses of Nainital (Himalayas, India). - *J. Hattori Bot. Lab.* No. **23**: 80-84.

Chopra, R.N. & Vijaykanta 1966. Some Mosses from Delhi, India. - *Bryologist* **69**(4) : 509-510.

Chopra, R.S. 1975. Taxonomy of Indian Mosses. CSIR Publication, New Delhi.

Chopra, R.S. 1977. New combinations in Indian mosses. - *The Bryologist* **80**: 543-544.

Chopra, R.S., B.K. Abrol, & M.S. Banga 1956. A preliminary list of the Mosses of Mussoorie Region. - *Res. Bull. Panjab Univ. Sci.* No. **85**: 21-30.

Chopra, R.S., S.S. Kumar & A.R. Dhand 1973. Preliminary report of the mosses of Simla. - *Nova Hedwigia.* **47**: 23-33.

Dabhade, G.T. 1998. *Mosses of Khandala & Mahabaleshwar in Western Ghats (India).* Thane, India.

Daniels, A.E.D. & P. Daniel 2003. *Fissidens griffithii* Gangulee (Musci : Fissidentales) - an addition to the Bryoflora of India. - *Indian J. Forestry* **26**(2): 193-194.

Daniels, A.E.D. & P. Daniel 2003. Additions to the Bryoflora of Peninsular India. - *Indian J. Forestry* **26**(4): 389-396.

Deb, D.B. 1955. Mosses of Manipur. - *Bull. Bot. Soc. Bengal* **9**: 81-83.

Dixon, H.N. 1909. Mosses from Western India. - *J. Bombay Nat. Hist. Soc.* **19**: 536-537.

Dixon, H.N. 1909. Mosses from Western Ghats. - *J. Bot., London* **47**: 157-164.

Dixon, H.N. 1910. *Merceyopsis*, a new genus of moss, with further contribution to the bryology of India. - *J. Bot. London* **48**: 297-310.

Dixon, H.N. 1911. *Hyophilopsis*, a new genus of Pottiaceae with further contributions to the bryology of India. - *J. Bot. London* **49**: 137-150.

Dixon, H.N. 1912. *Bryosedwickia*, novum genus Entodontacearum with further contributions to the bryology of India - *J. Bot. London* **50**: 145-146.

Dixon, H.N. 1914. Report on the mosses of the Abor Expedition 1911-12, and report on the mosses collected by Mr. C.E.C. Fisher and others from South India, Ceylon. - *Rec. Bot. Surv. India* **6**(3): 55-73; 75-89.

Dixon, H.N. 1921. On a collection of mosses from Kanara district. - *J. Indian Bot.* **2**: 174-188.

Dixon, H.N. 1925. Mosses from the Mt. Everest expedition, 1924. - *J. Bot. London* **63**: 221-222.

Dixon, H.N. 1926. Mosses collected in Gilgit, etc. by J. Garrett and W. Lillie. - *Rec. Bot. Surv. India* **9**(5): 303-313.

Dixon, H.N. 1930. Additions to the moss flora of North-western Himalayas. - *Annls Bryol.* **3** : 51-70.

Dixon, H.N. 1937. Mosses collected in Assam. - *J. Bombay Nat. Hist. Soc.* **39**(4): 769-795.

Dixon, H.N. 1942. Some new species of Acrocarpous Mosses from North-West Himalayan Flora. - 150th Anniv. Vol. R. Bot. Gdn. Calcutta: 173-182.

Dixon, H.N. & R.L. Badhwar 1938. Some new North-West Himalayan Mosses. - *Rec. Bot. Surv. India* **12**(2): 163-179.

Dixon, H. N. & Potier de la Varde, R. 1928. Musci Novi Indici -*Annls Cryptog.* exot. **1**: 37-47.

Dixon, H.N. & Potier de la Varde, R. 1928. *Homaliopsis* Dix. & Potier de la Varde, R. Generae Novi Muscorum. -*Annls Bryol.* **1**: 47-48.

Dixon, H.N. & Potier de la Varde, R. 1930. Nouvelle contribution a la flore bryologique de l' Inde. - *Annls Cryptog. exot.* **3**: 168-193.

Duthie, J.F. 1895. The Botany of Chitral Expedition. - *Rec. Bot. Surv. India* **1**: 139-181. (Mosses named by Brotherus. 180-181).

Fernandez, J. 1930. A list of mosses from Darjeeling district. - *J. Bombay Nat. Hist. Soc.* **44**: 600-601.

Foreau, G. 1930-1931. Notes on Bryological Geography for the Presidency of Madras. - *J. Madras Univ.* **2**: 238-250, 1930; **3**: 118-126. 1931.

Foreau, G. 1961. Moss Flora of Palni Hills. - *J. Bombay Nat. Hist. Soc.* **58**(1): 13-47.

Gangulee H.C. 1969-1980. *Mosses of Eastern India & Adjacent regions.* Fascicles 1-8. Calcutta.

Greven, H.C. 2000. *Grimmia indica* (Grimmiaceae), a new combination.- *J. Bryol.* **22**: 141.

Griffith, W. 1842. Muscologia Itineris Assamici (or a description of mosses collected during the journey in Assam, 1835-1836). - *Calcutta J. Nat. Hist.* **2**: 65-512.

Griffith, W. 1949. Notulae ad Plantas Asiaticas, Calcutta. Mosses. 386-485.

Griffith, W. 1949. Icones Plantarum Asiaticum. II. Calcutta.

Gupta, K.M. 1933. On the structure of a new species of Indian mosses. *Physcomitrellopsis indica* Dix. *sp. nov.* from Benaras. - *J. Indian Bot. Soc.* **12**: 122-128.

Handoo, O.N. & Sudha Srivastava 1964. *Nanomitrium tenerarum* (Bruch) Lindb. a new record from India. - *Curr. Sci.* **33**(5): 148-149.

Harvey, W.H. & J.D. Hooker 1840. Musci Indici. List of mosses collected in East India by Dr. Wallich, with reference to the figures of the new species by Hon. W.H. Harvey in Hooker's Icones Plantarum Vol. I. Tables XVII-XXIV (1840) and to which are added those collected by Dr. Royle in the Northern part of India by J.D. Hooker. - *J. Bot. (Hooker) & Kew Gdn Misc.* **2**: 1-21.

Hedwig, J. 1801. Species muscorum frondosorum descriptio et tabulis aensis 77 colaratis illustrata. Leipzig.

Herzog, T. 1939. Zwei Bryophytens ammulgen aus dem Sikkim- Himalaya. - *Annls Bryol.* **12**: 86-97.

Hooker, W.J. 1808. Musci Nepalensis. - *Trans. Linn. Soc. Lond. (Bot.)* **9**: 307-372.

Iwatsuki, Z & A.J. Sharp 1976. Notes on *Rhachithecium* in Assam, India. *Misc. Bryol. Lichenol.* **7**: 105-107.

Kaul, R.K. & G.C. Singh 1972. Some mosses from Kashmir. - *Bryologist* **75**(4): 586-588.

Kumar, S.S. & C.N. Negi 1977. Additions to the Moss Flora of Himachal Pradesh - I - *Bryologist* **80**(1): 206-207.

Kumar, S.S., Prem Lal Uniyal & Anju Goyal 1988. Additions to the Moss flora of Garhwal Himalaya II. - *Yushania* **5**(1): 19-23.

Kumar, S.S., Sarita Ganju & Sunita Ganju 1979. Additions to the Moss flora of Himachal Pradesh -II. *Nova Hedwigia* **63**: 339-342.

Kumar, S.S., Sunita Anand & Prem Lal Uniyal 1987. A report on the moss flora of Garhwal Himalaya- I. Research Bulletin (Science) of the Panjab University, Vol. 38. Parts III-IV, P: 39-41.

Lal, Jagdish 1985. *Brachymenium lanceolatum* Hook. *f. et* Wils. (Musci: Bryaceae) new to Indian Bryoflora. - *Yushania* **2**(3): 1-2.

Lal, Jagdish 2000. *Distichophyllum mittenii* Bosch et Lac. (Musci : Hookeriaceae) new to Indian bryoflora. - *Recent Trends in Botanical Researches* (ed. D.K. Chauhan) : 77-79. Botany Dept., Allahabad Univ.

Lal, Jagdish 2004. *Mosses of Central Indian Zone.* In Press.

Levier, E. 1907. A list of Himalayan mosses for sale without description.- *Revue Bryol.* **34**: 55.

Linnaeus, C. 1753. *Species Plantarum* ed. 2. vol. II. Musci. Holmiae.

Manju, C. Nair *et al.* 2004. *Bryum tuberosum* Mohamed & Damanhuri, a new record for India. - *Indian J. Forestry* **27**(1): 39-40.

Mitten, W. 1859. Musci Indiae Orientalis. - *J. Linn. Soc. Bot. Suppl.* **1**: 1-171.

Mueller, C. 1853. Musci Neilgherienses I. - Bot. Ztg. Regensburg **11**: 17-21; 33-40; 57-62.

Mueller, C. 1854. Musci Neilgherienses II. - Bot. Ztg. Regensburg **12**: 556-559; 569-574.

Noguchi, A. 1958. A small collection of Mosses from southern India. - *Kumamoto J. Sci. ser.* B. sec. 2, **4**: 1-4.

Noguchi, A. 1964. A list of mosses from Kashmir and Pakistan. - *Candollea* **19:** 191-198.

Noguchi, A., N. Takaki & H. Inoue 1966. Bryophytes collected by Dr. K. Yoda in Eastern Nepal. - *Bull. Nat. Sci. Mus. Tokyo* **9**(3): 360-378.

Noguchi, A., H. Suzuki, T. Asada, N. Takaki, Z. Iwatsuki, H. Ochi & U. Mizushima 1966a. Musci in the flora of the Eastern Himalayas. (Results of the Botanical expeditions to Eastern Himalayas organised by Univ. of Tokyo, 1960-1963, compiled by Hiroshi Hara Univ. of Tokyo, Japan. 537- 591.

Noguchi, A. 1971. Musci in the flora of the Eastern Himalayas. Second report compiled by Hiroshi Hara, Univ. of Tokyo, Japan: 241-258.

Norkett, A.H. 1958. Himalayan moss notes. I. *Orthomnium.* - *Trans. Br. Bryol. Soc.* **3**: 441-447.

Norkett, A.H. 1962. *Fissidens minutulus* Sull. and *F. pussillus* (Wils.) Milde. - *Trans. Br. Bryol. Soc.* **4**: 201-203.

Ochi, H. 1956. Supplement to the North western Himalayan Mosses. - *Bryologist* **59**: 213-216.

Ochi, H. 1963. Notes on Asiatic Moss Flora I. - Advancing Frontiers of Plant Sciences. **4**: 105-126.

Ochi, H. 1964. Notes on Asiatic Moss Flora. II. Mosses of the families Bryaceae and Bartramiaceae collected by members of the Univ. of Tokyo, Botanical Expedition 1960. - *J. Jap. Bot.* **39**: 49-59.

Ochi, H. 1968. Supplement to the mosses of the family Bryaceae collected by the members of the Univ. of Tokyo, Botanical Expd. to Eastern India 1963. - *J. Jap. Bot.* **43**: 109-114.

Pant G. & S.D. Tewari 1982. *Leucobryum cucullifolium* Card. and *L. nilghiriense* C. Muell. new records for North-West Himalayas. - *Indian J. Forestry* **5**(4): 320-321.

Pant, G. & S.D. Tewari 1984. *Distichophyllum schmidtii* Broth. and *Thamnobryum latifolium* (Bosch & Lac.) Nieuwl. -New records for India. - *Indian J. of Forestry* 7(2): 159-160.

Pant, G., S.D. Tewari & M.C. Pargaien 1986. *Homaliodendron scalpellifolium* (Mitt.) Fleisch., a new record for western Himalayan moss flora.- *Yushania* **3**(2): 35-37.

Parihar, N.S., Lal, Jagdish & Madhu Raj 1975. *Macrocoma* (Hornsch. *ex* C. Muell.) Grout (Orthotrichaceae) - a taxon new to India. - *J. Bryology* **8**: 475-478.

Potier de la Varde, R. 1922-1924. Musci Madurenses. I. - *Revue Bryol.* **49**(3): 33-44. 1922.

Potier de la Varde, R. 1925. Nouvelles herborisation dans le sud de I' Inde - *Revue Bryol.* **52**: 37-43.

Potier de la Varde, R. 1928. Mousses nouvelles de la Inde meridionale - *Annls Cryptog. exot.* **1**: 27-281.

Potier de la Varde, R. 1928a. Musci novii Indiai -*Annls Cryptog. exot.* **1** : 37-47.

Potier de la Varde, R. 1930. Sur deux mousses Indiennes. - *Annls. Cryptog. exot.* **3** (2-3): 101-105.

Reese W.D. & B.C. Tan 1986. *Syrrhopodon burmensis*, the correct name for *Syrrhopodon robustiusculus* (Musci : Calymperaceae). *Taxon* **35**(4): 693-694.

Robinson, H. 1964. A small collection of Bryophytes from upper Assam, India. - *J. Hattori Bot. Lab.* **27**: 124-130.

Robinson, H. 1965. A small collection of Bryophytes from Kashmir - *Bryologist* **68**: 313-320.

Robinson, H. 1968. Notes on the Bryophytes from the Himalayas and Assam-*Bryologist* **77**(2): 82-97

Sande Lacoste, C.M. Van der 1866. Species novae vel Ineditae Muscorum Archipelagi Indici. -*Verh. K. Akad. Wet. Amsterdom* **13**: 1-12.

Sharma, A.K. 1949. Indian Sphagnums. - *Bull. Bot. Soc. Bengal* **3**: 99-111.

Sharma, S., V. Nath & A.K. Asthana 2002. *Brachythecium garovaglioides* Müll. Hal., an addition to West Himalayan Bryoflora. - *Cryptogamie, Bryologie* **23**(3): 221-224.

Smith, G.L. 1974. Three new Himalayan species of *Polytrichastrum*. - *J. Hattori Bot. Lab.* **38**: 633-638.

Srivastava, J.P. 1966. Mosses of Kumaon 1. - *Proc. nat. Acad. Sci. India.* Sect. B, **36**: 385-396.

Srivastava, J.P. 1985. Studies on the mosses of Kumaon Himalaya. New Taxa -1. *Mielichhobryum* Srivastava (*gen. nov.*) - *Nat. Acad. Sci. Letters,* **8**(1): 15-18.

Srivastava, J.P. 1989. Mosses of Kumaon Himalayan New Taxa - III *Mnium kumaonensis* a new species. - *Nat. Acad. Sci. Letters* Vol. **12**(6): 177-179.

Srivastava, J.P. & A.H. Norkett 1988. *Fissidens arunii* Srivastava & Norkett *sp. nov.* - *Proc. Bot. Soc. Kanpur (India).* **1** : 55-57.

Stern, R.C. 2000. Some bryophytes collected in Rajasthan. - *J. Bryol.* **22**: 298-299.

Takaki, N. 1966. Bryophytes collected by Dr. K. Yoda in Eastern Nepal. - *Bull. Nat. Sci. Mus.* **9**(3): 359-386.

Tewari, S.D., B.S. Kalakoti, G.S. Rawat & G.B. Pant 1982. An enumeration of Pindari mosses. - *Himalayan Research & Development* **1**(II): 197-198.

Tewari, S.D. & G. Pant 1987. *Syrrhopodon spiculosus* Hook. & Grev.- a new record for India. - *J. Indian Bot. Soc.* **66**: 432-433.

Tewari, S.D. & G. Pant 1989. *Splachnobryum procerrimum* Dix. & P. Vard., a little known species from Western Himalaya (India). - *J. Bryol.* **15**: 555-557.

Tewari, S. D. & G. Pant 1990. *Splachnobryum synoicum* Robins. on new to the western Himalayas (India). - *J. Bryol.* **16**: 122-124.

Tewari, S.D., G. Pant, S.K. Airi & P.S. Adhikari 1987. *Hookeria acutifolia* Hook. & Grev. - a new record for western Himalaya, India.- *Yushania* **4**(3): 13-16.

Udar, R. *et al.* 1970. Genus *Buxbaumia* in India. - *Curr. Sci.* **39**(1): 14-15.

Udar, R. *et al.* 1971. A new species of *Buxbaumia* Hedwig., *B. himalayensis* Udar *et al.* from Deoban, Western Himalayas India. - *Trans. Br. Bryol. Soc.* **6**(2): 266-269.

Vashistha B.D. & R.N. Chopra 1982. *Regmatodon declinatus* (Hook.) Brid. - a new record for the North-western Himalaya India. - *Curr. Sci.* **51**(22): 1075-1076.

Vashistha, B.D. & R.N. Chopra 1983. Additions to the moss flora of North-western India. - *Miscellanea Bryologica et Lichenologica* **9**(8): 177-179.

Vitt, D.H. 1984. Classification of Bryopsida. -in R.M. Schuster (ed.) New Manual of Bryology. **2**: 696-759. Nichinan, Japan.

Vohra, J.N. 1966. *Merceyopsis angulosa* Broth. et Dix. -A new record for Western Himalayas. - *Bull. Bot. Surv. India* **8**: 355-356.

Vohra, J.N. 1967. A new species of *Hydrogonium* from the Western Himalayas- *J. Bombay Nat. Hist. Soc.* **63**(2): 464.

Vohra, J.N. 1969. New records of mosses from Western Himalayas. - *Bull. Bot. Surv. India* **10**(3,4): 394-396.

Vohra, J.N. 1971a. A contribution to the moss flora of Western Himalayas. *Ibid* **11**(1,2): 41-50.

Vohra, J.N. 1971b. A new species of *Habrodon* and some unrecorded mosses from India. *Ibid* **11**(1,2): 70-75.

Vohra, J.N. 1972. Mosses from Kashmir. *Ibid* **11**(3,4): 301-310.

Vohra, J.N. 1983. *Hypnobryales suborder Leskeineae of the Himalayas* - Records of the Botanical Survey of India **23**: 1-336.

Vohra, J.N. & B.D. Kar 1996. On a collection of mosses from Arunachal Pradesh. - *Bull. Bot. Surv. India* **38**(1-4): 46-54.

Vohra, J.N. & B.D. Kar 1996. On a collection of mosses from Great Nicobar Island. - *Bull. Bot. Surv. India* **38**(1-4): 55-59.

Vohra, J.N. & B.M. Wadhwa 1963. *Andreaea rupestris* Hedw. - a new record from the Western Himalayas - *J. Bombay Nat. Hist. Soc.* **5**: 149-150.

Vohra, J.N. & B.M. Wadhwa 1964. Mosses collected during Nilkanth and Chawkhamba expedition 1959. - *Bull. Bot. Surv. India* **6**(1): 43-46.

Vohra, J.N. *et al.* 1982. *Botanical Studies* on *Silent Valley* Part -I. Observation on the cryptogamic flora of Silent Valley. Mosses: 3-7; 19-30. BSI Publication. Howrah.

Wadhwa, B.M. & J.N. Vohra 1963. Name change for *Didymodon obtusifolius* Card. *ex* Dix. *et* P. Vard. - *Curr. Sci.* **32**: 482-483.

Wadhwa, B.M. & J.N. Vohra 1965. Miscellaneous Notes. - *J. Bombay Nat. Hist. Soc.* **62**(1): 177-179.

Wallich, N. 1828-49. A numerical list of dried specimens of plants in the East India Company's Museum (Calcutta) collected under supervision of Dr. Wallich. Mosses are Nos. 7545-7582.

Williams, R.S. 1932. Two undescribed Himalayan Mosses. - *Bryologist* **35**: 1-5.

Wilson, W. & W. Mitten 1857. Enumeration of the mosses collected in India by Dr. J.D. Hooker and Dr. T. Thomson, with their habitats, elevations and numbers under which they have been distributed, arranged by W. Mitten Esq. and named by W. Wilson Esq. - *J. Bot. (Hooker) & Kew Gdn Misc.* **9**: 289-300; 321-333; 363-370.

Zander, Richard H. 1993. *Genera of the Pottiaceae: Mosses of Harsh Environments* - Bull. Buffalo Society of Natural Sciences, Buffalo, N.Y. **32**: 1-378.